# AL-IDRISI'S NORMAN KINGDOM IN THE SOUTH

## FOUNDATIONS

**Further Information and Publications**

https://www.arc-humanities.org/series/book-series

# AL-IDRISI'S NORMAN KINGDOM IN THE SOUTH

## *THE BOOK OF ROGER* IN TRANSLATION

Edited and translated by

KATHERINE JACKA,
AHMED FATIMA KZZO,
MATT KING, and
SHERIF ABDELKARIM

ARC HUMANITIES PRESS

**British Library Cataloguing in Publication Data**
A catalogue record for this book is available from the British Library.



ISBN (hardback): 9781802700190
e-ISBN (PDF): 9781802701951

**www.arc-humanities.org**

Printed and bound in the UK (by CPI Group [UK] Ltd), USA (by Bookmasters), and elsewhere using print-on-demand technology.

# CONTENTS



*KITAB NUZHAT AL-MUSHTAQ FI IKHTIRAQ AL-AFAQ*

THE BOOK OF PLEASURE FOR THOSE
WHO DESIRE TO VISIT FARAWAY PLACES

# LIST OF ILLUSTRATIONS

## Figures



## Maps

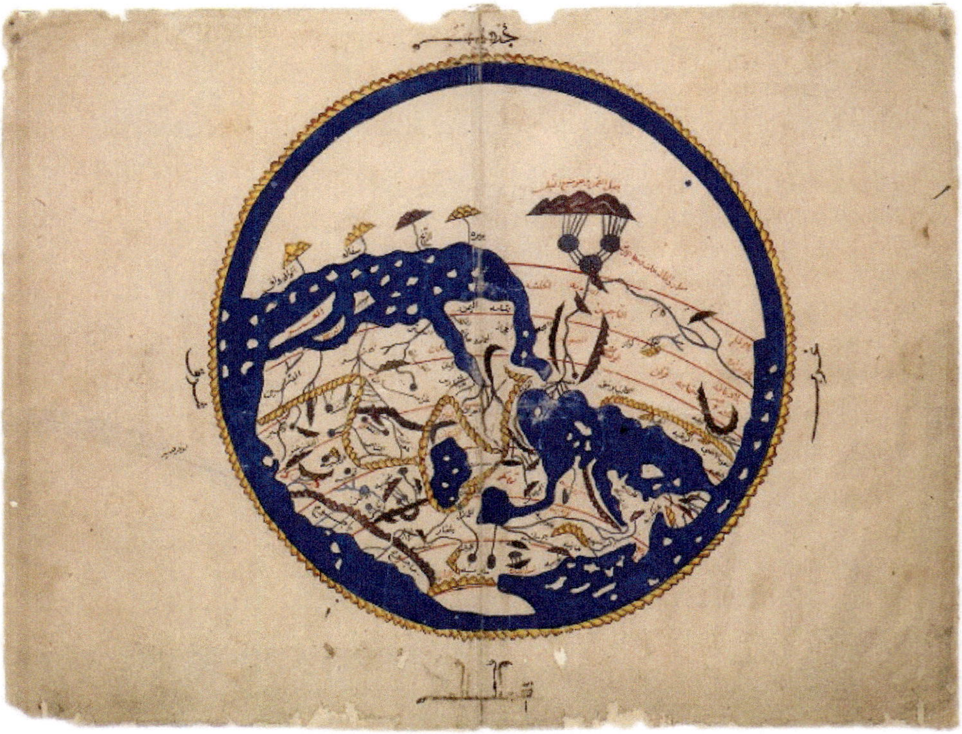

Figure 1. World Map (Oxford, Bodleian Library, MS Pococke 375, fols. 3b–4a).

Introduction

# THE LIFE, TIMES, AND WORKS OF AL-IDRISI

IN THE MIDDLE of the twelfth century CE, a Muslim geographer based in Sicily named Abu ʿAbdallah Muhammad al-Idrisi (d. 1164/5–1175) penned an expansive description of the world. The title of this work, *Kitab Nuzhat al-Mushtaq fi Ikhtiraq al-Afaq* (*The Book of Pleasure for Those who Desire to Visit Faraway Places*), gives only a small sense of its scale. This geography describes the people, settlements, flora, fauna, and geography of lands stretching from sub-Saharan Africa to Scandinavia and from Iberia to eastern Asia. It was an ambitious attempt at sketching out the entirety of the *oikumene*—a Greek geographical term meaning "inhabited earth."[1] Today, this geography is also known by a Latinized name, *Tabula Rogeriana* (*The Book of Roger*), which elevates the man who sponsored it: the Norman king of Sicily, Roger II (1095–1154). Across this monograph, we will refer to this text as the *Nuzhat al-Mushtaq* so that the Arabic-language origins of this text remain front and centre.

The *Nuzhat al-Mushtaq* of al-Idrisi was not only a sprawling written description of the known world. It also contained a number of maps: seventy regional maps organized into a 7 × 10 grid and one large world map showcasing the entirety of the known globe.[2] Indeed, these maps often serve as an accessible introduction to the work of al-Idrisi (and the Middle Ages more generally) thanks to their frequent appearances on book covers and other medieval-themed media.[3] While this cartographic art provides a useful point of entry to the *Nuzhat al-Mushtaq*, too often these maps serve as substitutes for the expansive writings of al-Idrisi, which contain far more information about human settlement patterns, economic trends, and forms of cultural expression. Our translation seeks to provide an accessible introduction to some of the chapters found within the *Nuzhat al-Mushtaq*, namely those regions that were governed by the

---

* The authors would like to thank Anna Henderson of Arc Humanities Press for her longstanding support of this project. In addition, we would like to thank Nizar F. Hermes for his generosity in providing advice on our translation.

1 Klaus Geus, "Oikoumene/Orbis Terrarum," in *Oxford Classical Dictionary*, 2016, https://doi.org/10.1093/acrefore/9780199381135.013.8008, accessed November 21, 2023.

2 It is unclear if the world map was included in the original manuscript of al-Idrisi, as he does not mention it within his preface. Jean-Charles Ducène, "Al-Idrīsī, Abū ʿAbdallāh," in *Encyclopaedia of Islam* (Leiden: Brill, 2018), http://dx.doi.org/10.1163/1573-3912_ei3_COM_32372, accessed November 21, 2023. Henceforth, all entries from the *Encyclopaedia of Islam* will be abbreviated as "[Entry]," *EI[# of edition]*.

3 See, for example, Sarah Davis-Secord, *Where Three Worlds Met: Sicily in the Early Medieval Mediterranean* (Ithaca: Cornell University Press, 2017); Matt King, *Dynasties Intertwined: The Zirids of Ifriqiya and the Normans of Sicily* (Ithaca: Cornell University Press, 2022). See also a digital exhibit from the Bibliothèque nationale de France titled "al-Idrîsî: la Méditerranée au XIIe siècle." Accessible at http://classes.bnf.fr/idrisi/, accessed November 21, 2023.

Normans in Ifriqiya (roughly modern Tunisia), Sicily, and south Italy. We hope that this translation will expand the audience of those able to appreciate the work of al-Idrisi not merely for the visual art accompanying it, but also for the text within it.

This introduction provides readers with information about the life, times, and work of al-Idrisi such that our subsequent translation can be read with an eye toward its historical context. An analysis of the physical texts of the *Nuzhat al-Mushtaq* that survive to the present day is an ideal place to begin.

## Manuscripts and Translations of the *Nuzhat al-Mushtaq*

Ten manuscripts of the *Nuzhat al-Mushtaq* exist in European and Middle Eastern libraries. None of these manuscripts were produced by al-Idrisi or were even written during his lifetime. All ten are copies made at least a century after the text was completed and vary greatly in quality.[4] It is likely that most of these manuscripts originated in North Africa—including the Bibliothèque nationale de France's MS Arabe 2221 and 2222, the Süleymaniye Library in Istanbul's *Ayasofya* 3502 and *Gugrafiya* 705 and the Bodleian's MS Pococke 375 and MS Greaves 42—and were copied between the thirteenth and sixteenth centuries before being transported to other libraries.[5] Of these texts, MS Arabe 2222 from the Bibliothèque nationale de France in Paris provided the foundation for later translations in the nineteenth and twentieth centuries.

Despite the number of surviving manuscripts of the *Nuzhat al-Mushtaq*, for most of the twentieth and early twenty-first centuries, scholars were reluctant to use it as a source for twelfth-century Sicily and the Mediterranean due in part to the "poor reputation of existing editions and translations."[6] An abridged Arabic version of the *Nuzhat al-Mushtaq* published by the Medici Oriental Press in Rome in 1592 has rightly been maligned by modern scholars as an inadequate abridgement created "apparently without any precise motive."[7] The Medici version of the *Nuzhat al-Mushtaq* was later translated into Italian by a scholar and theologian based at the Scuola di Urbino, Bernadino Baldi, around 1600, though the utility of this translation is constrained by the

---

**4** For an exhaustive list of extant manuscripts, see S. Maqbul Ahmad, "Cartography of Al-Sharīf al-Idrīsī," in *Cartography in the Traditional Islamic and South Asian Societies*, ed. J. B. Harley and David Woodward, The History of Cartography 2.1 (Chicago: University of Chicago Press, 1992), appendix 7.1.

**5** The maps from the Bodleian Library are perhaps the most well-known of this selection. They are accessible at al-Idrisi, *Bodleian Library MS. Pococke 375*, 1553, https://digital.bodleian.ox.ac.uk/objects/ced0d8bd-1019-4af2-9086-e411115f1507/surfaces/cc2debab-70f5-4b82-9bab-7235fdb89e43/, accessed November 21, 2023. The North African provenance of these manuscripts is due to the family of Sharafi al-Sifaqsi (based in the city of Sfax in modern Tunisia), which created their own maps using the geography of al-Idrisi as their model. Ahmad, "Cartography of Al-Sharīf al-Idrīsī," 170.

**6** David Abulafia, "Local Trade Networks in Medieval Sicily: The Evidence of Idrisi," in *Shipping, Trade and Crusade in the Medieval Mediterranean: Studies in Honour of John Pryor*, ed. Ruthy Gertwagen and Elizabeth Jeffreys (Farnham: Ashgate, 2012), 157.

**7** "Al-Idrīsī," *EI2*. See also Alberto Tinto, *La tipografia Medicea Orientale* (Lucca: Pacini Fazzi, 1987).

flawed abridgement from which it derives.[8] The same is true of a Latin version based on the Medici text that was produced in Paris in 1619, which was translated by two Maronite monks under the title of *Geographia Nubiensis*.[9]

The only translation of the entire *Nuzhat al-Mushtaq* is a French one completed between 1836 and 1840 by the French Arabist Pierre Amédée Jaubert, based largely on the Bibliothèque nationale de France's MS 2222, which contains the full text of al-Idrisi's work but no maps.[10] Although Jaubert's efforts are considerable, the translation has not aged particularly well. When Henri Bresc and Annliese Nef published in 1999 a new revision of the portions of the *Nuzhat al-Mushtaq* that describe the Mediterranean, they remarked that the work of Jaubert contains "a number of inaccuracies and gaps, as well as misunderstandings of the text of al-Idrisi."[11] The partial translation of Bresc and Nef is the most recent attempt at translating a large portion of the *Nuzhat al-Mushtaq*, and though it is a wonderful work of scholarship, it is of little help to those without reading knowledge of French who wish to understand this medieval work.

Scholars have produced a number of other partial translations of the *Nuzhat al-Mushtaq*, most of which are in Italian and French. Michele Amari's foundational 1880 *Biblioteca arabo-sicula*, for example, provides an Italian translation of the *Nuzhat al-Mushtaq*'s preface and description of Sicily (based again on MS 2222).[12] Amari and Celestino Schiaparelli also published an Italian translation of al-Idrisi's description of the entire Italian peninsula in 1883, *L'Italia descritta nel "Libro del Re Ruggero"*, which was revized and updated in 1966 by Umberto Rizzitano.[13] An 1866 French partial translation by Dozy and Goeje, which was utilized by Bresc and Nef in their 1999 translation, is another admirable attempt at making the Arabic work of al-Idrisi accessible for a French-language audience.[14] The translation we have produced for this work, though based ultimately in the Arabic text of the *Nuzhat al-Mushtaq*, has benefitted from the translations and accompanying footnotes from these manifold works.

The substantial number of French and Italian translations of the *Nuzhat al-Mushtaq* does not mean that the Anglophone world has been entirely neglected. Scholars

---

**8** On the life of Bernadino Baldi, see Giovanni Ferraro, *Bernardino Baldi e il recupero del pensiero tecnico-scientifico dell'antichità* (Alessandria: dell'Orso, 2008).

**9** This title stems from a mistranslation of the Arabic text in which a passage dealing with Nubia mentions a river passing through "our countries" as opposed to "those" countries.

**10** Al-Idrisi, *Géographie d'Édrisi*, trans. Amédée Jaubert, 2 vols., Recueil de Voyages et de Mémoires 5 (Paris: Libraire de la Societé, 1836).

**11** Al-Idrisi, *La Première géographie de l'Occident*, trans. Henri Bresc and Annliese Nef (Paris: Flammarion, 1999).

**12** Michele Amari, *Biblioteca Arabo-Sicula, Versione Italiana*, vol. 1 (Torino: Ermanno Loescher, 1880), 31–133. Amari had previously published these portion of the *Nuzhat al-Mushtaq* in Arabic in Michele Amari, *Biblioteca Arabo-Sicula* (Leipzig: F.A. Brockhaus, 1857), 14–73.

**13** al-Idrisi, *L'Italia descritta nel "Libro del Re Ruggero"*, trans. Michele Amari and Celestino Schiaparelli (Rome: Tipi del Salviucci, 1883); al-Idrisi, *Il Libro di Ruggero*, trans. Umberto Rizzitano (Palermo: Flaccovio, 1966).

**14** al-Idrisi, *Description de l'Afrique et de l'Espagne Par Edrisi*, trans. Reinhart Dozy and Michael Jan Goeje (Leiden: Brill, 1866).

have published a handful partial English translations of the *Nuzhat al-Mushtaq* over the last century. In 1950, for example, Beeston published an article that included a critical transcription and translation of al-Idrisi's sections on the British Isles.[15] Mones similarly translated the portion of the *Nuzhat al-Mushtaq* pertaining to Egypt in a 1984 article published in *Studi Magrebini*.[16] Portions of the *Nuzhat al-Mushtaq* that consider Saharan and sub-Saharan West Africa are translated, too, in Levtzion and Hopkins' *Corpus of Early Arabic Sources for West African History*.[17] Graham Loud and Alex Metcalfe also translated part of the preface and Sicily section from al-Idrisi for the *Roger II and the Creation of the Kingdom of Sicily*.[18] These combined texts, though useful in their own right, represent only a fraction of the overall scope of the *Nuzhat al-Mushtaq*, and they pale in comparison to the number of texts translated into French and Italian. We hope that our translation will build on the work of these previous English-language scholars to include coverage of areas not systematically considered, namely those areas of the Mediterranean under Norman control in the central Mediterranean when al-Idrisi penned his geography.

Our translation is based on the Arabic critical edition of the book published in nine fascicles between 1970 and 1982 under the auspices of the *Istituto Italiano per il Medio ed Estremo Oriente*, with an editorial committee based at the *Istituto Universitario Orientale di Napoli*. This critical edition was undertaken by prominent Italian Arabists, including Roberto Rubinacci and (the abovementioned) Umberto Rizzitano, and bears the Latin title, *Opus Geographicum, Sive "Liber Ad Eorum Delectationem Qui Terras Peragrare Studeant"*.[19] This edition represents the most comprehensive and recent critical edition of the *Nuzhat al-Mushtaq* and remains the standard Arabic-language version of al-Idrisi's geography. In rare cases where we deviate from the body of the text within this edition, we discuss this choice in accompanying footnotes.

## The Normans in the South

When al-Idrisi wrote the *Nuzhat al-Mushtaq* in the middle of the twelfth century, he was serving at the royal court of King Roger II in Palermo. There, he was part of a group of administrators, scholars, and artists whose synthetic productions elevated

**15** A. F. L. Beeston, "Idrisi's Account of the British Isles," *Bulletin of the School of Oriental and African Studies* 13 (1950): 265–80.

**16** Hussain Mones, "Description of Egypt by Idrisi," *Studi Magrebini* 16 (1984): 1–53.

**17** Nehemia Levtzion and J. F. P. Hopkins, eds., *Corpus of Early Arabic Sources for West African History*, trans. J. F. P. Hopkins (Princeton: Markus Wiener, 2000), 104–31.

**18** Loud initially translated the French translation published by Bresc and Nef in 1999. Alex Metcalfe then checked that English translation against the original Arabic text of al-Idrisi. Graham Loud, *Roger II and the Creation of the Kingdom of Sicily* (Manchester: Manchester University Press, 2012), 355. Karla Mallette also published a short excerpt from the preface of the *Nuzhat al-Mushtaq* in Karla Mallette, *The Kingdom of Sicily, 1100–1250: A Literary History* (Philadelphia: University of Pennsylvania Press, 2005), 146–48.

**19** Al-Idrisi, *Opus geographicum, sive "Liber ad eorum delectationem qui terras peragrare studeant"*, ed. Enrico Cerulli, 9 vols. (Leiden: Brill, 1970).

the Norman regime by sponsoring works that appropriated the stylings of other Mediterranean powers. The process through which it became possible for a Muslim scholar to hold a privileged position in the capital of a Christian king, especially during an era when crusading fever was sweeping Europe, is a complex one that deserves some explanation.

Around the beginning of the eleventh century, a band of pilgrims from the duchy of Normandy in northern France made a pilgrimage to the Church of the Holy Sepulcher in Jerusalem. On their way back from the Holy Land, they stayed with a Lombard lord in south Italy and helped him repel enemies at his doorstep (either Muslims or Byzantines depending on the source).[20] In subsequent decades, waves of additional Norman soldiers poured into Italy—first ostensibly to assist local lords, and then to carve out territory for themselves. South Italy at the time of these conquests was politically fragmented, as Lombard and Byzantine lords controlled patches of land while papal and German interests spurred further divisions in the region. The Normans took advantage of these parochial rivalries to gain large tracts of land for themselves. By the middle of the eleventh century, the Normans had established themselves as independent lords in fortified settlements across the Italian peninsula, especially in the regions of Apulia and Calabria. Some of the most prominent individuals to participate in these early conquests were sons of the Norman noble Tancred de Hauteville, including William "Iron Arm" and his brother Drogo.

Norman elites in Italy were an extreme minority on the Italian peninsula, so they relied on local allies and reinforcements from northern France to bolster their nascent rule.[21] Many Normans married into Lombard families and used their combined military might to force out Byzantine lords from the region. The ascendency of Norman rule in south Italy brought the Normans many enemies: the Byzantines that they defeated, the Lombards whose territory they partitioned, the Papacy at their doorstep to the north, and local inhabitants who suffered from years of conquest and accompanying destruction. John of Fécamp, a Benedictine monk, wrote a letter to Pope Leo IX in the mid-eleventh century in which he described how "this hatred of the Italians toward the Normans has flared up and now grown so much that through all of the regions of Italy it is scarcely possible for any Norman (unless he is a devoted pilgrim) not to be seized and jumped, dragged away, stripped [of his belongings], beaten, bound in

**20** On this history, see David Abulafia, *The Two Italies: Economic Relations between the Norman Kingdom of Sicily and the Northern Communes* (New York: Cambridge University Press, 1977); Graham Loud, *The Age of Robert Guiscard: Southern Italy and the Norman Conquest* (New York: Longman, 2000); Joanna Drell, *Kinship and Conquest: Family Strategies in the Principality of Salerno during the Norman Period, 1077–1194* (Ithaca: Cornell University Press, 2002); Graham Loud, "Southern Italy in the Eleventh Century," in *The New Cambridge Medieval History*, 7 vols., vol. 4.2 (New York: Cambridge University Press, 2004), 94–119.

**21** Issues surrounding Norman identity are explored in Alex Metcalfe, *The Muslims of Medieval Italy* (Edinburgh: Edinburgh University Press, 2009); David Bates and Pierre Baudin, eds., *911–2011: Penser les mondes normands médiévaux* (Caen: Presses universitaires de Caen, 2016); Judith A. Green, *The Normans: Power, Conquest & Culture in 11th-Century Europe* (New Haven: Yale University Press, 2022).

chains" and left to prison, where many died.[22] Although the Normans received some degree of papal legitimacy after winning a victory against German, Lombard, and papal forces at the Battle of Civitate in 1053, their presence on the peninsula remained contested and (in many cases) reviled.

As the Normans pressed their advantages across the Italian peninsula, a cohort of Normans looked to expand their domains west into Sicily—the largest island in the Mediterranean Sea. Like south Italy, Sicily was politically divided when the Normans first set foot on the island in 1061.[23] A handful of Muslim powers vied for control of the island, including the remains of the Kalbid dynasty, princes from the Zirid dynasty of Ifriqiya, and numerous other independent lords (especially on the eastern edge of the island). Norman forces under the brothers Robert Guiscard and Roger Bosso seized on this fragmentation by making an alliance with Ibn al-Thumna against a local rival, which secured them a foothold on the island. Over the next thirty years, Norman forces slowly subdued the lords of Sicily through a series of sieges and pitched battles, including several victories over Zirid forces based in western Sicily. The Normans conquered Palermo, the largest city on the island and its political capital, in 1072. By 1091, the entire island was under Norman control, though there remained pockets of mountainous inland Sicily where it was difficult for the Normans to exercise their authority.

The death of Robert Guiscard in 1085 meant that his brother Roger was able to assume control of Norman Sicily without provoking infighting from his relatives. As Count of Sicily, Roger ruled over an island with sprawling commercial connections, agricultural wealth, and diverse inhabitants.[24] Prior to the arrival of the Normans, Sicily had been a crucial node in trading networks that traversed the Mediterranean. The exchange of Sicilian goods in nearby Ifriqiya and Egypt was a particularly important network, as these ports had once been part of the sprawling Fatimid Empire and were well equipped to handle maritime commercial traffic. Sicilian durum wheat, for example, was hardier than other contemporary varieties and could be sold at high prices in foreign markets on the southern shores of the Mediterranean. Traders in Sicily used

---

**22** John of Fécamp, "Epistola Joannis I Abbatis Fiscamnensis ad S. Leonem IX," in *Patrologia Latina*, ed. J. P. Migne, vol. 143 (Paris: Migne, 1853), 797–800.

**23** On Sicily prior to the Normans see Alex Metcalfe, *Muslims and Christians in Norman Sicily: Arabic Speakers and the End of Islam* (New York: Routledge Curzon, 2003); Annliese Nef, *Conquérir et gouverner la Sicile islamique aux XIe et XIIe siècles* (Rome: École française de Rome, 2011); Glaire D. Anderson, Corisande Fenwick, and Mariam Rosser-Owen, eds., *The Aghlabids and Their Neighbors: Art and Material Culture in Ninth-Century North Africa* (Leiden: Brill, 2017); David Bramoullé, *Les Fatimides et La Mer (909–1171)* (Leiden: Brill, 2019).

**24** On these trading networks, see S. D. Goitein, "Sicily and Southern Italy in the Cairo Geniza Documents," *Archivio Storico per La Sicilia Orientale* 67, no. 1 (1971): 9–33; Sarah Davis-Secord, "Sicily and the Medieval Mediterranean: Communication Networks and Inter-Regional Exchange" (PhD diss., University of Notre Dame, 2007); Annliese Nef, "La Sicile dans la documentation de la Geniza Cairote (fin xe-xiiie): les réseaux attestés et leur nature," in *Espaces et réseaux en Méditerrannée (VIe–XVIe siècle)*, vol. 1, *La configuration des réseaux*, ed. Damien Coulon, Christophe Picard, and Dominique Valérian (Saint-Denis: Bouchène, 2007), 273–92.

this wealth to purchase silk, spices, salt, and gold (among other goods). Although the Norman conquest brought some degree of disruption to the trade networks that circulated through Sicily, they persisted throughout the eleventh century and into the reign of Roger I.[25]

When Roger I died, in 1101, he was succeeded by his young son, Roger II.[26] The new count and his mother, Adelaide, managed to secure his rule on the island by suppressing a handful of small revolts.[27] With their possessions secure, Roger II was able to weaponize his expansive demesne lands and the wealth of Sicily to his advantage.[28] He expanded the bureaucracy of the Norman court at Palermo (considered in more detail below) and aggressively pursued Norman lands in south Italy. Following the death of William II, Duke of Apulia and Calabria, in 1127 Roger gradually conquered all of the lands in south Italy south of the Papal States. These conquests brought him into direct confrontation with the papacy again, though another Norman victory over papal forces led to Roger II being formally recognized as King of Sicily in 1139. In the

**25** Evidence for the Cairo Geniza is crucial for understanding these patterns. See, for example, S. D. Goitein, *Letters of Medieval Jewish Traders* (Princeton: Princeton University Press, 1973); Ruthy Gertwagen, "Geniza Letters: Maritime Difficulties along the Alexandria-Palermo Route," in *Communication in the Jewish Diaspora: The Pre-Modern World*, ed. Sophia Menache (Leiden: Brill, 1996), 73–90; Shlomo Simonsohn, *The Jews in Sicily: 383–1300* (Leiden: Brill, 1997); Shlomo Simonsohn, *Between Scylla and Charybdis: The Jews in Sicily* (Leiden: Brill, 2011).

**26** Latin sources provide the most detail on the life of Roger II. They include Hugo Falcandus, *La Historia o Liber de Regno Sicilie e la Epistola ad Petrum Panormitane Ecclesie Thesaurarium*, ed. G. B. Siragusa, Fonti per la Storia d'Italia (Rome: Tipografi del Senato, 1897); Romuald of Salerno. *Romualdi Salernitani Chronicon*, edited by Carlo Garufi. Rerum Italicarum Scriptores 7.1 (Citti di Castello: S. Lapi, 1935); Geoffrey Malaterra, *De rebus gestis Rogerii Calabriae et Siciliae comitis et Roberti Guiscardi ducis fratris eius*, ed. L. A. Muratori and Ernersto Pontieri, Rerum Italicarum Scriptores 5 (Bologna: Zanichelli, 1949); Alexander of Telese, *Alexandri Telesini Abbatis Ystoria Rogerii Regis Sicilie Calabrie atque Apulie*, ed. Ludovica de Nava, Fonti per la Storia d'Italia 112 (Rome: Instituto Palazzo Borromini, 1991); Falco of Benevento, *Chronicle Beneventanum: città e feudi nell'Italia dei Normanni*, ed. Edoardo D'Angelo (Florence: Sismel, 1998). Translations of these texts can be found in Hugo Falcandus, *The History of the Tyrants of Sicily by "Hugo Falcandus," 1154–69*, trans. Graham Loud and Thomas Wiedemann (Manchester: Manchester University Press, 1998); Geoffrey Malaterra, *The Deeds of Count Roger of Calabria and Sicily and of His Brother Duke Robert Guiscard*, trans. Kenneth Baxter Wolf (Ann Arbor: University of Michigan Press, 2005); Loud, *Roger II and the Creation of the Kingdom of Sicily*.

**27** Hubert Houben, "Adelaide 'del Vasto' nella storia del regno normanno di Sicilia," in *Mezzogiorno normanno-svevo. Moansteri e castella, ebrei e musselmani*, ed. Hubert Houben (Naples: Liguori, 1996), 81–115; Joshua Birk, *Norman Kings of Sicily and the Rise of the Anti-Islamic Critique: Baptized Sultans* (New York: Palgrave Macmillan, 2016), 81–138.

**28** There is a substantial body of modern scholarship about the reign of Roger II. See, for example, David Abulafia, "The Crown and the Economy under Roger II and His Successors," *Dumbarton Oaks Papers* 37 (1983): 1–14; Hubert Houben, *Roger II of Sicily: A Ruler between East and West*, trans. Graham Loud and Diane Milburn (New York: Cambridge University Press, 2002); Julia Becker, "Charters and Chancery under Roger I and Roger II," in *Norman Tradition and Transcultural Heritage: Exchange of Cultures in the "Norman" Peripheries of Medieval Europe*, ed. Stefan Burkhardt and Thomas Foerster (New York: Ashgate, 2013), 79–96; Dawn Marie Hayes, *Roger II of Sicily: Family, Faith, and Empire in the Medieval Mediterranean World* (Turnhout: Brepols, 2020).

decades that followed, Roger expanded the borders of his kingdom through attacks on Byzantine territories in the eastern Mediterranean and Muslim lands in North Africa (considered later in this Introduction).

Like the Normans in Italy, those in Sicily were minorities on the island. Roger I had relied upon military conquest, marriage alliances, and generous donations to churches to solidify his control over the various groups in Sicily. Although most living on the island were Muslim, a substantial minority of Greek Christians existed on the eastern side of the island. Roger, who had experienced firsthand the animosity between the Normans and Greek Byzantines in south Italy, looked to elevate Latin Christian worship in his lands by donating land and money to Latin churches. At the same time, though, Roger and his successors adapted Byzantine regnal propaganda for their own rule. Roger II, for example, is portrayed in a contemporary mosaic from the Church of Santa Maria dell'Ammiraglio (Palermo) as being crowned by Christ himself in a pose that is reminiscent of earlier Byzantine precedent. The ivory plaque of Byzantine Emperor Constantine VII Porphyrogenitus provides a useful juxtaposition for this particular case of Norman adaptation.[29]

The Normans also appropriated bureaucratic structures and visual propaganda from the Muslim world to bolster their regime. The Normans moved their Sicilian capital to Palermo, which had been the seat of power of the Muslim Kalbid dynasty, during the regency of Adelaide in the early twelfth century. Palermo and the lands surrounding it were inhabited primarily by Muslims, and the Normans sought to legitimize their rule by emulating aspects of other Muslim dynasties from around the Mediterranean.[30] The defection of George of Antioch, a Greek Christian administrator previously employed by the Zirids of Ifriqiya, to Palermo in the early twelfth century gave the Normans a wealth of knowledge of the inner workings of Zirid/Fatimid governance.[31] The Normans struck coins with Arabic titles and imported artisans to emulate art from Fatimid Egypt—seen today most strikingly in the ceiling of the Cappella Palatina in Palermo.[32] The Normans also began producing government documents written in

**29** Norman propaganda is considered in Karen Britt, "Roger II of Sicily: Rex, Basileus, and Khalif? Identity, Politics, and Propaganda in the Cappella Palatina," *Mediterranean Studies* 16 (2007): 21–45; Jeremy Johns, "The Quadrilingual Epitaph of Anna, the Mother of Grisandus, a Priest in the Cappella Palatina, Palermo," in *The Visual Culture of Later Byzantium (c. 1081–c. 1350)*, ed. Foteini Spingou, Sources for Byzantine Art History 3 (Cambridge: Cambridge University Press, 2021), 2:1448–61.

**30** Jeremy Johns, "The Norman Kings of Sicily and the Fatimid Caliphate," *Anglo-Norman Studies* 15 (1993): 133–59; Jeremy Johns, *Arabic Administration in Norman Sicily: The Royal Dīwān* (New York: Cambridge University Press, 2002).

**31** Johns, *Arabic Administration in Norman Sicily*, 80–90; Adalgisa de Simone, "Note sui titoli arabi di Giorgio di Antiochia," in *Byzantino-Sicula V: Giorgio di Antiochia: l'arte della politica in Sicilia nel XII secolo tra Bisanzio e l'Islam*, ed. Christina Rognoni and Mario Re (Palermo: Istituto siciliano di studi bizantini e neoellenici Bruno Lavagnini, 2009), 283–308.

**32** Jeremy Johns, "I titoli arabi dei signori normanni di Sicilia," *Bollettino di numismatica* 6–7 (1986): 11–54; Philip Grierson and Lucia Travaini, *Italy III: South Italy, Sicily, Sardinia*, Medieval European Coinage 14 (New York: Cambridge University Press, 1998), 179–96.

Arabic and adopted elements of Fatimid rule to their image.[33] According to the Mamluk historian, al-Maqrizi, George of Antioch was instrumental to these reforms and worked to cultivate a regnal image of Roger II that was befitting of a Fatimid sultan:

> [George] amassed [Roger's] revenues and organized the foundations of the kingdom. He veiled Roger from [his] subjects, and arranged for him to dress in clothes like the Muslims, and not to ride out, nor to show himself in public, except on holidays, when he would process, preceded by horses adorned with saddles of gold and silver...Roger's state grew under George's management... Thus the island flourished in a most splendid way, and travellers from every land made for it with all sorts of goods and rare merchandise.[34]

The wealth of the Norman court at Sicily afforded Roger II and his successors the opportunity to patronise art and literature. Roger financed the importation of classical Greek works from Constantinople (including Plato's *Meno* and *Phaedo*) and had them translated into Latin.[35] He also kidnapped silk workers from Thebes and had them create works of regnal propaganda that simultaneously showcased the various cultures present in Sicily but also the domination of the Normans over them.[36] This idea is on full display in the coronation mantle of Roger II, which contains Kufic-style Arabic text listing the accomplishments of Roger II but simultaneously shows lions devouring camels—a clear allusion to the Norman conquest of Muslim lands.[37] It is in this milieu of syncretic propaganda and tenuous religious tolerance that al-Idrisi composed his geography, and it is to his life and work that we now turn.

## A Murky Biography of al-Idrisi

Although a substantial corpus of evidence exists to document the history and culture of Norman Sicily, we know relatively little about the life of Muhammad al-Idrisi.[38] Arabic sources provide the most details about him, though they are rarely in agree-

**33** These reforms outlived George of Antioch, who died in 1151/52. When the Muslim traveler Ibn Jubayr passed through the island of Sicily during the reign of William II (r. 1166–89, grandson of Roger II), he praised the Norman regime for its tolerance and adoption of Muslim cultural mores. Ibn Jubayr, *Rihla Ibn Jubayr* (Beirut: Dar Sadir, 1964); Ibn Jubayr, *The Travels of Ibn Jubayr: A Medieval Journey from Cordoba to Jerusalem*, ed. Robert Irwin, trans. Ronald Broadhurst (New York: I. B. Tauris, 2019).

**34** Translation is from Johns, *Arabic Administration in Norman Sicily*, 80–82. Al-Maqrizi, *Kitab al-Muqaffa al-Kabir*, ed. Muhammad al-Yaʿlawi, 8 vols. (Beirut: Dar al-Gharb al-Islami, 1991).

**35** Henricus Aristippus, *Phaedo interprete Henrico Aristippo*, ed. Laurentius Minio-Paluello (London: Warburg Institute, 1950).

**36** David Jacoby, "Silk Crosses the Mediterranean," in *Le Vie Del Mediterraneo: Idee, Uomini, Oggetti (Secoli XI–XVI)*, ed. G. Airaldi (Genoa: Ed. culturali internazionale genova, 1997), 55–79.

**37** Isabelle Dolezalek, *Arabic Script on Christian Kings: Textile Inscriptions on Royal Garments from Norman Sicily* (Berlin: De Gruyter, 2017).

**38** The most accessible and thorough biography of al-Idrisi in English comes from Jean-Charles Ducène in his entry for the third edition of the *Encylopaedia of Islam*; "al-Idrīsī, Abū ʿAbdallāh," *EI3*.

ment. In the biographical dictionary *al-Wafi bi-l-Wafayat* (Sufficiency of obituaries), the Syrian biographer al-Safadi (ca. 1297–1363) details al-Idrisi's life and his connection to Sicily through his entries on King Roger II, al-Idrisi himself, and al-Idrisi's grandfather.[39] Al-Safadi records that al-Idrisi descended from the Moroccan Idrisid dynasty and the Hammudids of Málaga, dynasties that could trace their lineage back to the Prophet Muhammad himself. Al-Idrisi was likely educated in Córdoba, the intellectual centre of Andalusia, as indicated by the *nisba* (an adjective used in Arabic names to indicate a person's birthplace, tribe, or ancestry) of "al-Qurṭubi" ("the Cordoban").[40] Nonetheless, the place of al-Idrisi's birth and location of his upbringing is disputed. It is possible that he lived much of his early life in either North Africa, Sicily, or south Italy. Al-Safadi provides contradictory accounts on this point. In his entry on al-Idrisi himself, al-Safadi writes that al-Idrisi "grew up" (*nashåa*) in Sicily: "This Muḥammad grew up amongst the companions of Roger the Frank, Ruler of Sicily. He was a writer, a witty poet, and captivated by the science of geography."[41]
However, in a later entry on King Roger, al-Safadi writes: It was he who invited al-Sharif al-Idrisi, the author of *Nuzhat al-Mushtaq fi Ikhtiraq al-Afaq* to come from the *ʿUdwa* in order to set down something about the form of the world.[42] When he reached him, [Roger II] honoured [al-Idrisi] and showered him with veneration.[43] Adding to this uncertainty, al-Safadi's entry on al-Idrisi includes a few poems (purportedly) penned by al-Idrisi himself.[44] These poems have dark undertones and deal with issues of identity, belonging, and nostalgia for a distant homeland—potentially indicating that al-Idrisi was in exile at Roger's court in Palermo:

> I've known intimately peoples and lands
> Both good and evil,

---

**39** al-Safadi, *Kitab al-Wafi bi-l-Wafayat*, ed. Ahmad al-Arnåut and Turki Mustafa, 29 vols. (Beirut: Dar Ihyå al-Turath al-ʿArabi, 2000). On this entry, see Tarek Kahlaoui, *Creating the Mediterranean: Maps and the Islamic Imagination* (Leiden: Brill, 2017), 143–45.

**40** Al-Idrisi's full name is Abu ʿAbdallah Muhammad al-Idrisi al-Qurtubi b. Muhammad b. ʿAbdallah b. Idris b. Yahya al-ʿAli bi-Amrillah b. ʿAli b. Hammud al-ʿAli. "al- Idrīsī, Abū ʿAbdallāh," *EI3*.

**41** al-Safadi, *Kitab al-Wafi bi-l-Wafayat*, 1:138–39.

**42** The meaning of *ʿudwa* (shore/coast/bank) is open to interpretation. Allaoua and Nef argue that *ʿudwa* could refer to the Calabrian coast or Sicily. At the same time, though, the term might reference the *Barr al-ʿUdwa*—part of North Africa that includes Ceuta. Al-ʿUmari, *Al-Taʿrīf Bi-al-Mustalah al-Sharif* (Beirut: Dar al-Kotob al-Ilmiyah, 1988); Allaoua Amara and Annliese Nef, "Al-Idrisi et les Hammudides de Sicile: Nouvelles donnees biographiques sur l'auteur du livre de Roger," *Arabica* 48 (2001): 121–27; Annliese Nef, "Al-Idrīsī : un complément d'enquête biographique," in *Géographes et voyageurs au Moyen Âge*, ed. Henri Bresc and Emmanuelle Tixier du Mesnil (Nanterre: Presses universitaires de Paris Nanterre, 2010), 53–66.

**43** al-Safadi, *Kitab al-Wafi bi-l-Wafayat*, 14:72–73.

**44** For poems with similar thematic ideas, see William Granara, "Ibn Hamdis's al-Dimas Qasida: Memorial to a Fallen Homeland," in *Poetry and History: The Value of Poetry in Reconstructing Arab History*, ed. Ramzi Baalbaki, Saleh Said Agha, and Tarif Khalidi (Beirut: American University of Beirut Press, 2011), 247–64; William Granara, *Ibn Hamdis the Sicilian: Eulogist for a Falling Homeland* (London: Oneworld, 2021).

Yet found no neighbour, no home
As inheres within my heart.
It is as though I do not roam,
With neither corpse nor grave.[45]

Although the circumstances of al-Idrisi's early life are unclear, his career comes into (slightly) sharper focus by the 1130s. As the preface to the Nuzhat al-Mushtaq indicates, King Roger II was interested in scientific and geographical works but was unimpressed with current offerings. He therefore tasked al-Idrisi with creating a more satisfying geography, which took at least fifteen years to complete and involved the writing not only of the *Nuzhat al-Mushtaq* but also the creation of a large silver globe of the world (the content of which is considered shortly). The date of the completion of the *Nuzhat al-Mushtaq* is contested. Scholars long thought that it was completed in 1154 (the year of Roger II's death), but several of the events mentioned in the work—such as the mention of Frederick Barbarossa living in Burgundy—indicate that it (or parts of it) were completed later in the 1150s.[46] The *Nuzhat al-Mushtaq* was not the only work that al-Idrisi wrote in Palermo. He also penned a compendium on drugs and a second, shorter, book of geography that is today commonly called the "Little Idrisi."[47]

Perhaps unsurprisingly, the later life and death of al-Idrisi is also disputed. The reign of William I of Sicily saw the rise of anti-Muslim violence in Sicily, which perhaps motived al-Idrisi and some of his co-religionists to emigrate from the island. Leo Africanus states that he died in Ceuta in 1164/65 (perhaps adding credence to the idea that this was the location of his birth as well), while ʿImad al-Din al-Isfahani claims that he died in Ifriqiya in 1175.[48] We are fortunate to have numerous manuscript copies of the *Nuzhat al-Mushtaq* survive to this day, as the large silver globe of the known world that accompanied it was destroyed in 1161 during a revolt against Roger II's successor, William I.

## The Influences and Content of the *Nuzhat al-Mushtaq*

Although al-Idrisi's geography was formed under the auspices of a Christian king, it drew heavily on a rich lineage of Arabic geographical literature, *al-Masalik wa-l-Mamalik* (Routes and kingdoms), which dates back to the court of the Abbasid caliph al-Mȧmun (r. 813–833). These texts, which adapted Greek and Sasanian precedents, emerged from an Abbasid effort to govern a large, culturally heterogeneous, and quickly expanding empire—a situation not unlike that of Roger II. Under the Abbasid

**45** al-Safadi, *Kitab al-Wafi bi-l-Wafayat*, 1:138–39.

**46** "al-Idrīsī, Abū ʿAbdallāh," *EI3*.

**47** On the compendium of drugs, see Christina La Rosa, "La Terminologia Botanica Sicula e Andalusa nel Kitāb al-Ḏjāmiʿ li-as̱ẖtāt al-nabāt di Idrīsī," *Annali di Ca' Foscari* 50 (2014): 97–121. On the "Little Idrisi," see Jean-Charles Ducène, *L'Afrique dans le Uns al-Muhağ Wa-Rawḍ al-Furağ d'al-Idrīsī* (Leuven: Peeters, 2010); Jean-Charles Ducène, "Les œuvres géographiques d'al-Idrīsī et leur diffusion," *Journal Asiatique* 305 (2017): 33–41.

**48** al-Idrīsī, Abū ʿAbdallāh," *EI3*.

caliphs a new multicultural society had developed in Iraq, based on the diverse demographic mix of the region, altering the cultural orientations of the Arab empire drastically.[49] The Abbasids were at pains to appear as the natural successors to the ancient imperial dynasties of Mesopotamia and Iran, and as such they enthusiastically incorporated Indo-Persian, and later Hellenistic, culture into their own. The monastic libraries of conquered cities contained original Pahlavi scholarship as well as Greek and Syriac work of scholars such as Plato, Aristotle, Hippocrates, Galen, and, importantly for the later work of al-Idrisi, Claudius Ptolemy.[50]

Drawing on Ptolemy's *Almagest*, al-Maʿmun commissioned his astronomers and geographers to depict the earth's surface by creating a world map, or perhaps more accurately, an atlas.[51] Although influenced by Ptolemy, this was not to be simply a reconstruction of Ptolemy's world view but a completely new work which took into consideration the political and geographical conditions of the burgeoning Islamic empire. This work came to be known as *al-Ṣurat al-Maʿmuniya* (al-Maʿmun's map) and represented a substantial development on from Ptolemy, adding two new zones south of the equator as well as correcting Ptolemy's depiction of the Indian Ocean as a land-locked sea. Although there is no extant copy of the map, Masʿudi wrote a century after the map was produced that it presented "the world with its spheres, stars, land, and seas, the inhabited and uninhabited regions, settlements of peoples and cities and other such things. This was better than what had come before it, either the Geography of Ptolemy, the Geography of Marinus, or any other."[52] The map included some 530 important cities and towns as well as seas, rivers, and mountains as well as information on "famous and marvellous things" (*ʿajāʾib*) within these territories.[53]

Due to the work of the cartographers of the eighth and ninth centuries, a corpus of reliable maps had been created, and in the tenth century there was a shift in the focus of geographical scholarship: now there was a move to descriptive, or human, geography which came to be known as Routes and Kingdoms. Works in the Routes and Kingdoms genre described human settlement patterns and included information like distances between settlements, ports, and anchorages, freshwater sources,

**49** Dimitri Gutas, *Greek Thought, Arabic Culture. The Graeco-Arabic Translation Movement in Baghdad and Early 'Abbasid Society* (London: Routledge, 1998), 18–21.

**50** Amira K. Bennison, *The Great Caliphs: The Golden Age of the Abbasid Empire* (New Haven: Yale University Press, 2009), 175. For more on Arabic translations of Ptolemy's *Almagest* and *Geography* see Paul Kunitzsch, "The Arabic Translations of Ptolemy's Almagest," *Qatar Digital Library* www.qdl.qa/en/arabic-translations-ptolemys-almagest, accessed November 21, 2023; and Jean-Charles Ducène, "Ptolemy's Geography in the Arabic-Islamic Context" in *Cartography between Christian Europe and the Arabic-Islamic World, 1100–1500* (Leiden: Brill, 2021).

**51** Fuat Sezgin, *Mathematical Geography and Cartography in Islam and Their Continuation in the Occident*, (Frankfurt am Main: Institute for the History of Arabic-Islamic Science, 2000–2007), 2:78.

**52** Masʿudi, *Al-Tanbih wa l-ishraf* in *Bibliotheca Geographorum Arabicorum*, ed. Michael Jan de Goeje, vol. 8 (Leiden: Brill, 1892; reprinted 1967 and 2014), 33.

**53** Quotation from the twelfth-century geographer from Grenada, ʿAbu ʿAbdallah al-Zuhri in Jonathan Lyons, *The House of Wisdom: How the Arabs Transformed Western Civilisation* (London: Bloomsbury, 2009), 89.

commercial goods, population, cultural production, and defensive structures. These entries had clear utility for the Abbasid court, which could use this information to inform their taxation patterns, laws, and military priorities. Early writers in this genre like Ibn Khurdadhbih (820–912) focused on the religious and cultural centers of the Islamic world, especially Baghdad or Mecca, while peripheral areas like North Africa and Sicily received comparatively little attention (if they were included at all).[54]

In a bid to redress these lacunae, from 1000 CE original Routes and Kingdoms work began to appear in the western portion of the *Dar al-Islam*, in particular in Spain.[55] In this work precedence was given to the western places of production, with exhaustive research and information about Spain and North Africa, areas that had been neglected by scholars such as Ibn Hawqal, no doubt in order to downplay the impact of the resurgent Umayyads. The most prominent authors of this new western-based Muslim geography were all from Andalusia and like the Routes and Kingdoms scholars of the east, were attached to government, benefitting from access to government records. This geographical scholarship worked to feature Muslim Spain, giving it primacy, and the decision to exclude the rest of the *Dar al-Islam* was doubtless a calculated one. Prominent writers of this new western geography included al-Bakri (1040–1095) and al-ʿUdhri (1003–1085), and indeed al-Idrisi cites the latter as a source for the *Nuzhat al-Mushtaq*. Although he does not mention al-Bakri, it is probable al-Idrisi drew on al-Bakri's work especially in regard to the Maghreb, West Africa, and the Sudan. Given the fact that al-Idrisi had been educated in Córdoba, it is possible he had accessed the work of al-Bakri and al-ʿUdhri during his time in Spain.

Likewise in Abbasid Iraq, there was a very clear connection between Routes and Kingdoms writing and contemporary political conditions. The authors of these works were typically members of the Abbasid government and relied on the state to finance their work. These works thus acted as expressions of Abbasid power, and the territory that they considered "mattered...it stimulated the geographical imagination and acted as a powerful vehicle for articulating desire, claiming authority, and establishing belonging."[56] This same idea applies to the creation of the *Nuzhat al-Mushtaq* in twelfth-century Sicily. Like the Abbasids, Roger II was the newly appointed ruler of a culturally heterogeneous state where diversity was both an opportunity and a challenge—and where his authority was constantly called into question. In order to solidify his authority, Roger co-opted diverse Islamic, Byzantine, and Latin traditions in order to forge a uniquely Sicilian royal identity and to bolster the burgeoning Norman

**54** "Ibn Khurradādhbih," *EI2*; "Geography in Arabic," *EI3*.

**55** Emmanuelle Tixier du Mesnil, *Géographes d'Al-Andalus: De l'inventaire d'un territoire à la construction d'une mémoire* (Paris: Publications de la Sorbonne, 2014), 182–83. See also Alfred Hiatt, "Geography at the Crossroads: The Nuzhat al-mushtāq fī ikhtirāq al-āfāq of al-Idrīsī" in *Cartography between Christian Europe and the Arabic-Islamic World, 1100–1500* (Brill, 2021).

**56** Zayde Antrim, *Routes and Realms: The Power of Place in the Early Islamic World* (New York: Oxford University Press, 2012), 1. See also Travis Zadeh, *Mapping Frontiers across Medieval Islam: Geography, Translation and the 'Abbasid Empire* (London: Bloomsbury, 2017) and Karen C. Pinto, *Medieval Islamic Maps: An Exploration* (Chicago: University of Chicago Press, 2019).

state. The decision to commission the *Nuzhat al-Mushtaq* was part of this process. Roger took an Arabic geographical genre with prestigious roots in the Muslim East and appropriated it for his own ends by ensuring its focus was on the seat of his own kingdom in the central Mediterranean.

The *Nuzhat al-Mushtaq* borrows other conventions from medieval Arabic geographical treatises and their Greek counterparts. Al-Idrisi divides the globe into "Climes" (*iqlīm*, pl. *aqālīm* in Arabic) by using a system that was deployed by the Greek polymath Ptolemy and was later adopted by Muslim scholars.[57] These Climes roughly correspond to our modern idea of lines of latitude beginning at the equator and moving northward; al-Idrisi and his contemporaries believed that southern lands were uninhabited due to extreme heat.[58] However, the physical division of the world within the *Nuzhat al-Mushtaq* innovates on earlier formulas, as al-Idrisi divides the world into seven latitudinal Climes and then, originally, further divides these into ten longitudinal sections, resulting in seventy zones and seventy corresponding maps.[59] The description of these zones is uneven. Those that consider lands furthest from the Mediterranean (like East Asia) are often only several pages in modern typeset, while those covering lands in the Mediterranean and Middle East can extend for dozens of pages. This disparity in content is understandable given al-Idrisi's access to previous Arabic geographies, which tended to focus on the Middle East, and eyewitness accounts from Sicily, which likely dealt with Mediterranean lands.

Regardless of the amount of space allotted to a given region, a crucial component of the *Nuzhat al-Mushtaq* is the measurement of the distance between settlements and the distance between settlements and major geographical phenomena. Al-Idrisi uses two major types of measurements to record relative locations. The first is the "mile," which scholars today call the "Arabic mile"; it measures approximately two kilometres (though the precise measurement is disputed).[60] The second is a measurement based on the time it takes to travel to a given place. Al-Idrisi often refers to a location being one day's journey or two day's journey from another location—rendered in Arabic as "*marḥala*." When describing Sicily, he specifies that a journey of one day corresponds to around 25 miles (meaning around 50 modern kilometres), though this specification likely was not meant to correspond to all of the entries for which al-Idrisi used the term *marḥala*.[61] In any case, the terminologies that al-Idrisi used to describe the distances between locations were based on eyewitnesses whose perceptions of

**57** Ptolemy, *Ptolemy's Geography: An Annotated Translation of the Theoretical Chapters*, trans. Alexander Jones and J. Lennart Berggren (Princeton: Princeton University Press, 2020).

**58** See p. 39.

**59** Gabriel Martinez-Gros, "La division du monde selon Idrîsî," in *Le partage du monde: Échanges et colonisation dans la Méditerranée médiévale*, ed. Michel Balard and Alain Ducellier (Paris: Éditions de la Sorbonne, 1988), 315–34.

**60** Edward S. Kennedy, "Mathematical Geography," in *The Encyclopedia of the History of Arabic Science*, ed. Roshdi Rashed (New York: Routledge, 1996), 187–88; Fuat Sezgin, *Mathematical Geography and Cartography in Islam and Their Continuation in the Occident*.

**61** See p. 57.

geography were undoubtedly different.[62] We therefore should be skeptical of trying to make definitive statements about the relative locations of settlements based on the hard numbers that al-Idrisi provides, though these numbers are nonetheless useful for understanding rough approximations and contemporary perceptions of the distances between settlements.

Although al-Idrisi neatly divides the world into seventy distinct zones, the territories over which the Normans ruled in the Mediterranean do not fit neatly into these categories. Our translation, which focuses on territories governed by Roger II and his successors, therefore does not always align neatly with the geographical scheme laid out in the *Nuzhat al-Mushtaq*. We have translated the entirety of al-Idrisi's introductory preface; Clime 3, Part 2 (Ifriqiya); Clime 4, Part 2 (Sicily and nearby islands); and Clime 5, Part 3 (south Italy and the Adriatic). We have also provided partial translations of sections that include territories under Norman rule, but also contain non-Norman regions: Clime 3, part 1 (eastern Maghreb); Clime 4, Part 3 (south Italy and the Adriatic coast); and Clime 5, Part 2 (south Italy).[63] This is by no means an ideal arrangement. We would have preferred to translate all sections related to the Mediterranean or—even better—the entirety of the *Nuzhat al-Mushtaq*, but limitations in book length and our own expertise meant that we had to limit the portions of the *Nuzhat al-Mushtaq* that we translated for this book. We hope that these sections nonetheless expand the audience for al-Idrisi's work and can perhaps inspire other scholars to undertake translations of other portions of this geography.

## Within the *Nuzhat al-Mushtaq*: The Preface

The preface of the *Nuzhat al-Mushtaq* provides a laudatory explanation of Roger II's motivations for commissioning this geography, the methodology used by al-Idrisi within it, and an overview of the text's content. It provides the most detailed account of the text's creation and is of considerable importance for understanding the circumstances in which al-Idrisi penned this geography. At the same time, though, we should also bear in mind how the circumstances of this text's creation necessarily influence its content. Al-Idrisi was living and working at the court of the Norman king, so he was obligated to write a praiseworthy dedication that was worthy of his patron. His unrestrained praise of the Norman monarchy within this section tells us more about how Roger II wanted to project his power and image than about how a Muslim like al-Idrisi living in the Norman court perceived his Christian surroundings.

---

**62** The fact that al-Idrisi alternates between the use of miles and days within chapters of the *Nuzhat al-Mushtaq* is evidence, too, of the multiple sources that he encountered when researching this geography.

**63** Our translation of Clime 3, Part 1 only includes a few cities that the Normans briefly conquered or attacked in western Ifriqiya/eastern Maghreb. Clime 4, Part 3 considers south Italy and part of the southern Adriatic. Our translation of Clime 5, Part 2 likewise only mentions a handful of locations that the Normans controlled in Italy.

Although the presence of an esteemed Muslim author like al-Idrisi in the court of a Christian monarch might appear peculiar at the surface, it was not a rare occurrence during the central Middle Ages (and the reverse is true as well).[64] In exchange for memorializing their patrons in prose and verse, premodern courtiers enjoyed financial security, physical protection, and societal prestige—all of which was contingent upon their continued flattery of this patron. Within the Islamic world, this praise was most often expressed in a genre known as *madḥ* or *madīḥ*, which flattered such royal sponsors. Works in this style extol the virtues of powerful men (and occasionally women) by emphasizing their intelligence, commitment to justice, courage, and modesty.[65] Al-Idrisi's adulation of Roger II is written in keeping with these literary norms and comes on the heels of the preface's paragraph-long *Basmalah* and *Hamdalah*, ornate acclaims to Allah—a standard start to Islamic literary works, past to present. Al-Idrisi adapts this precedent to the court of his patron, however. He does not mention the prophet Muhammad, for example, and instead provides invocations to Allah that would be better received in a Christian context.

Even a quick glance of the *Nuzhat al-Mushtaq*'s preface reveals the degree of *madīḥ*-worthy praise that al-Idrisi heaped upon King Roger and his achievements. He describes the king as "the exalted King Roger, empowered by Allah the self-powered, king of Sicily, Italy, Lombardy, and Calabria; the victorious Roman leader of the Nazarene faith, in that he excels the Byzantine king in the distribution, acquisition, and delivery of matters, ratifying and voiding as he pleases."[66] Al-Idrisi notes the expansive territories in Sicily and south Italy ruled by the Normans, which Roger fought to conquer and receive papal legitimacy for during the 1120s and 1130s. He also praises Roger as superior to the Rum (here meaning the Byzantines) and elevates him as a defender of Latin Christianity. This language is no coincidence. Al-Idrisi wrote this passage when the Normans were launching attacks in the eastern Mediterranean against Byzantine strongholds, which he sought to elevate in this panegyric.[67] He also mentions these campaigns elsewhere in the preface when he declares that Roger "subdued the necks of the tyrants among his fold far and near with what he possesses from well-equipped troops of ready numbers, and heavy fleets."[68] Here, al-Idrisi does not limit his praise to the campaigns of Roger against the Byzantines, but instead notes his many military victories to the east and west—likely referring to his campaigns in the eastern Mediterranean and in Ifriqiya (to be considered shortly).

Al-Idrisi also lauds the political acumen of his patron, who "dealt among his faithful justly, protecting them under the wing of grace and favour. He raised with the ropes of his rule an excellent effectuation: he handled the flow of his polity according to

**64** Luke B. Yarbrough, *Friends of the Emir: Non-Muslim State Officials in Premodern Islamic Thought* (Cambridge: Cambridge University Press, 2019).

**65** "Madīḥ, Madḥ," *EI2*.

**66** See p. 36.

**67** These campaigns are considered in Charles Stanton, *Norman Naval Operations in the Mediterranean* (Woodbridge: The Boydell Press, 2011), 89–102.

**68** See p. 36.

the best organization, the most beautiful society, and opened lands east and west."[69] This political savvy is similarly matched by his intellect.[70] Al-Idrisi writes that Roger "added to fine manners noble ancestry, and to beautiful deeds an excellent disposition, plus spiritual courage; mental purity; intellectual depth; abundant understanding; sound judgement and planning; expertise in the execution of affairs out of a complete, piercing perception."[71] These characteristics, though certainly overly grandiose and inflated in keeping with stylistic conventions of the genre, nonetheless echo much of what is found in pro-Norman Latin texts from this time.[72] Scholars, though, have a more complicated relationship with the personality and abilities of Roger II, who indeed succeeded at uniting fractured territories in Sicily and south Italy, but did so through acts of both generosity and extreme violence. When Roger died in 1154, his kingdom was at its greatest extent, but the numerous enemies that he had made when conducting these conquests meant that his son William I was stuck between multiple powers that sought to destroy his lands.[73] The intellectual curiosity of Roger II is also difficult to adjudicate. Although he sponsored many artistic and literary works at his court in Palermo, he did so in a manner reminiscent of other Mediterranean rulers such that he could legitimize and further the power of his own regime.

In keeping with this image of King Roger as an astute monarch, the preface of the *Nuzhat al-Mushtaq* presents his intellectual curiosity as central to the creation of this text. Al-Idrisi writes that Roger sought "to know the properties of his lands truly; to study them definitively and expertly; to know their limits and methods by land and sea" as well as those that lay within the seven Climes.[74] This curiosity of Roger II—likely the result of a desire to more security govern his lands and perhaps conquer those close to him (mixed with perhaps a dose of genuine curiosity)—motivated the king to seek additional information about the landscape around him. This drive

**69** See p. 36.

**70** Al-Idrisi even suggests that Roger was an inventor, who "contrived some unusual inventions and devised some strange devices" that "no kings have previously done", [page number forthcoming—typeset not determined]. It is unclear what creations (if any) al-Idrisi is referencing, though an 1142 trilingual inscription from the Cappella Palatina states that Roger had a water clock constructed. Houben, *Roger II of Sicily: A Ruler between East and West*, 104.

**71** [Page numbers forthcoming—typeset not determined].

**72** Alexander of Telese is perhaps the most praiseworthy. He writes, for example, "For if Vergil, the greatest of poets, was esteemed so much at the court of the Emperor Octavian that for two verses which he had written in his praise he received as an earthly reward rule over the city of Naples and at the same time the province of Calabria, we believe it much better that we should be rewarded by you with those things for which we ask for following the divine observance, not however so much for this little work as for the safety of your soul and for the Lord our Saviour, in whose name we are here gathered together, and by Whom you yourself have up to now been preserved safely and have achieved your kingdom." Translation is from Loud, *Roger II and the Creation of the Kingdom of Sicily*, 127.

**73** Graham Loud, "William the Bad or William the Unlucky?," *The Haskins Society Journal* 8 (1999): 99–113.

**74** See p. 37.

ultimately led to the formation of the *Nuzhat al-Mushtaq*. The process through which this geography was assembled, per the preface, revolved around the Norman king:

> [King Roger] summoned the experts of this matter, consulting and conversing with them about it. He then realized that they possessed no more knowledge than what was in the aforementioned books. When he saw them in this state, he called across his lands. The experts then attended in response, those esteemed in these studies. He asked them accordingly, collectively and individually, yet they arrived at no consensus. He approved the collection of their most resolved and enduring reports; whatever they disagreed upon he set aside or discarded. He maintained this procedure for about fifteen years, never depriving himself at any time from examining this field—its inquiry into it and research into its truth—until he attained what he desired from it.[75]

Per the preface, the importance of this project to Roger II was such that he ordered not only the production of a written geographical compendium, but also a large silver globe upon which major topographical features could be displayed. Al-Idrisi describes the completion of this monument and his orders to write a book to accompany it:

> [Roger] ordered that a purely silver sphere be made available to him...When it was completed, he ordered the labourers to engrave upon it pictures of the seven Climes with their lands; their shoreside and countryside; their gulfs, seas, and channels; their rivers' positions; its arable and barren areas; what lies between each of their lands and the others in terms of trodden paths and specified miles and attested distances and known ports, according to what comes to them—[all this] represented in a drawing map. They left nothing out of it, and produced it in its shape and look, as they saw it. And [he ordered them] to author a book that conforms to its forms and appearances, except that it add to the former by describing the conditions of the lands and terrains in their nature, location, positions; their shores, seas, mountains, distances; their crops, expenses, kinds of buildings; their properties; the tools they use; the industries that pay; the businesses that attract and export; the wonders said of and ascribed to them; where they fall within the seven Climes, with mention of the conditions of their people: their bearings, manners, orientations, fashions, clothing, and languages; and to name this book: *The Book of Pleasure for those who Desire to Visit Faraway Places*.[76]

The role of Roger II in the creation of the *Nuzhat al-Mushtaq* is almost certainly inflated within the preface, but beneath this grandiose portrait of an intellectual monarch lies important details about the formation of the *Nuzhat al-Mushtaq*.[77] The first is that it was a deeply collaborative enterprise based on eyewitness accounts. The amount of

---

**75** See p. 37.

**76** See p. 38.

**77** Kahlaoui, *Creating the Mediterranean*, 144–46.

original information found in the *Nuzhat al-Mushtaq* for territories controlled by the Normans speaks to the synthesizing of a substantial amount of original information that came from a diverse pool of sources. Even if Roger was not in charge of vetting all of this information himself, his desire for a contemporary and systematic geography spurred the collection of a significant amount of data about the central Mediterranean. It is unclear, though, who exactly these eyewitness sources were. Living in the port city of Palermo and sponsored by the royal court, al-Idrisi and other members of the Norman court had access to soldiers, pilgrims, government officials, and merchants who had traveled extensively throughout the Mediterranean, especially to those areas governed by the Normans in Ifriqiya, Sicily, and south Italy. These individuals, alongside firsthand experience of the author and information from existing geographies (which al-Idrisi mentions in the preface), provided the majority of geographical information from the *Nuzhat al-Mushtaq*. Across our translation, we have noted passages that bear striking resemblance to other Arabic geographies and, at times, considered how al-Idrisi might have known particular details about local landscapes.

The interest and potential involvement of the Norman monarch should also affect our reading of this text. Al-Idrisi was writing under the watchful eye of Norman officials who could easily oust him from his privileged position if his writings demeaned them, so he had to filter or camouflage any information that might paint the regime of Roger II in a negative light. The laudatory preface (in keeping with contemporary standards) elevates King Roger as a defender of Christianity, as a keen political operator, and as an intellectual. In the text of the *Nuzhat al-Mushtaq*, though, some entries are less praiseworthy of the Normans (especially in Ifriqiya—to be considered shortly), and these sections perhaps reveal some of the underlying tensions between the imperial ambitions of the Norman king and the feelings of the text's Muslim author. When reading the *Nuzhat al-Mushtaq*, we should be mindful of these messy historical intersections that make the work of al-Idrisi a text of both immense complexity and substantial historical value.

The remainder of the *Nuzhat al-Mushtaq*'s preface concerns the scientific methodologies that undergird the remainder of the text. Here, the influence of classical Greek texts is clear. Al-Idrisi cites Ptolemy early in this section and echoes many of geographical ideas found in Greek texts: the earth is round; life on extreme ends of the earth is impossible due to extreme temperatures; and the inhabitable part of earth is divided into seven Climes. Al-Idrisi further describes seven seas that "pierce" the seven Climes, which was a common idea in medieval Arabic geographical writing, and concludes with an overview for what readers can expect to encounter in the following seventy chapters.[78] The *Nuzhat al-Mushtaq* then launches into the first of its seventy chapters with an examination of west Africa. Within our limited translation, though, the first chronological chapter considers the landscape of Ifriqiya.

---

**78** Marina A. Tolmacheva, "Geography," in *Medieval Islamic Civilization: An Encyclopedia* (New York: Routledge, 2006).

## Within the *Nuzhat al-Mushtaq*: Ifriqiya

The description of Ifriqiya found in al-Idrisi's geography (Clime 3, Part 2) is built on both prolonged Norman interest in the region and, to a lesser extent, previous Arabic geographies.[79] When al-Idrisi began his geography in the late 1130s, the Normans had decades of involvement with Muslim lords in Ifriqiya—a relationship that was built on geographical proximity and historical precedent.[80] At their closest points, Sicily is separated from the coast of North Africa by less than one hundred miles. The waters between North Africa and Sicily are also dotted with numerous small islands that make it possible to navigate between the two landmasses without losing sight of land on a clear day.[81] The ease of transit between these two regions facilitated the exchange of goods across them. The Fatimid dynasty, which controlled both regions for long stretches of the tenth and eleventh centuries, forged strong commercial bonds between ports in Sicily, Ifriqiya, and Egypt that are evident today in letters from Jewish traders based in Cairo.[82] Central to the commercial relationship between Sicily and Ifriqiya during this time was the exchange of durum wheat from Sicily for Ifriqiyan gold (acquired from sub-Saharan Africa), salt, and slaves. Although other goods (like olive oil and textiles) were exchanged between these traders, the importation of grain to Ifriqiya was foundational to the relationship between the Normans and Muslim lords across the Strait of Sicily.

Economic trouble and military conflict in the eleventh century brought major changes to the political landscape of Ifriqiya, which had lasting consequences for the Normans in Sicily. The Zirid dynasty of Ifriqiya, which ruled nominally on behalf of the Fatimids of Egypt, suffered from recurrent droughts during the early 1000s that spurred emigration from Ifriqiya to Sicily and the Maghreb.[83] In 1008, a rival family

---

**79** Ifriqiya broadly corresponds to modern Tunisia, eastern Algeria, and western Libya. Its borders were based on the contours of the Roman province of *Africa Proconsularis* but changed over time based on the political dynamics of a given time. "Ifrīqiya," *EI3*; Ramzi Rouighi, *The Making of a Mediterranean Emirate: Ifriqiya and Its Andalusis, 1200–1400* (Philadelphia: University of Pennsylvania Press, 2011), 4–9.

**80** Parts of the analysis of the next two sections are adapted from Matt King and Katherine Jacka, "A GIS Mapping of Al-Idrisi's Geographies of Sicily and Ifriqiya in The Book of Roger," *The Medieval Globe* 9, no. 1 (2023): 27–83.

**81** Peregrine Horden and Nicholas Purcell, *The Corrupting Sea: A Study of Mediterranean History* (New York: Wiley-Blackwell, 2000), 127.

**82** Michel Balard, "Notes sur le commerce entre l'Italie et l'Égypte sous les Fatimides," in *L'Égypte fatimide: son art et son histoire* (Paris: Presses de l'Université de Paris-Sorbonne, 1999), 627–33; Yaacov Lev, "A Mediterranean Encounter: The Fatimids and Europe, Tenth to Twelfth Centuries," in *Shipping, Trade and Crusade in the Medieval Mediterranean : Studies in Honour of John Pryor*, ed. Ruthy Gertwagen and Elizabeth Jeffreys (New York: Taylor & Francis, 2012), 131–56; Michael Brett, *The Fatimid Empire* (Edinburgh: Edinburgh University Press, 2017).

**83** On the history of the Zirid dynasty, see H. R. Idris, *La Berbérie orientale sous les Zīrīdes: Xe–XIIe siècles*, 2 vols. (Limoges: A. Bontemps, 1962); Michael Brett, "Fatimid Historiography: A Case Study—The Quarrel with the Zirids, 1048–1058," in *Medieval Historical Writing in the Christian and Islamic Worlds*, ed. D. O. Morgan (London: School of Oriental and African Studies, 1982), 47–60;

broke off from the Zirids to form the Hammadid dynasty, which led to decades of on-and-off conflict. The most meaningful political change to occur during this century, though, was the arrival of the Banu Hilal from Egypt into Ifriqiya in the middle of the eleventh century. Initially, mercenaries that had migrated west from Egypt, the Banu Hilal slowly carved out territory for themselves in Ifriqiya and eventually displaced the Zirids and Hammadids from their inland capitals and to the coastal strongholds of Mahdia and Bougie (respectively). As factionalism within the ranks of the Banu Hilal emerged in subsequent decades, the political landscape of Ifriqiya became further fragmented, as the Zirids and Hammadids controlled patches of coastal land near their capitals and various Hilalian emirs occupied the remaining inhabitable lands.[84]

When the Normans conquered Sicily in the late eleventh century, their first encounters with the Zirids were on the battlefield. Once the conquest of Sicily was complete, though, both sides recognized that trade was more productive than conflict. Roger I signed a peace treaty with the Zirids around the late 1080s, and commerce persisted between Sicily and Ifriqiya in subsequent decades. This relationship continued into the reign of Roger II, but began to fracture in the 1110s as the Normans sought greater control over affairs in Ifriqiya. The Zirids were able to use regional alliances to repel the Normans several times during the 1120s, but a catastrophic drought in Ifriqiya during much of the 1140s decimated the population of the region, made Ifriqiyan lords even more reliant on increasingly unaffordable Norman grain, and thus provided an opportunity for the ambitious Normans. Led by George of Antioch, the Norman navy conquered many of the islands between Sicily and Ifriqiya, the Zirid capital of Mahdia, and a handful of other nearby population centres on the Ifriqiyan coast in the late 1140s. This so-called Norman Kingdom of Africa lasted for around a decade, as internal unrest from the Muslims of Ifriqiya and the ascendent Almohads destabilized Norman possessions in the region. The final Norman outpost in Ifriqiya, Mahdia, fell to the Almohads in 1160.[85]

---

Michael Brett, "The Diplomacy of Empire: Fatimids and Zirids, 990–1062," *Bulletin of the School of Oriental and African Studies* 78 (2015): 149–59.

**84** The arrival of the Banu Hilal into Ifriqiya has been the subject of substantial scholarly debate. See, for example, Jean Poncet, "Le mythe de la 'catastrophe' hilalienne," *Annales Histoire, Sciences Sociales* 22 (1967): 1099–1120; Roger Idris, "De la réalité de la catastrophe hilâlienne," *Annales Histoire, Sciences Sociales* 23 (1968): 390–96; Jean Poncet, "Encore à propos des hilaliens: la 'mise au point' de R. Idris," *Annales, Économies, Sociétés, Civilisations* 23 (1968): 660–62; Michael Brett, "The Zughba at Tripoli, 429H (1037–8 A.D.)," *Libyan Studies; Annual Report of the Society for Libyan Studies* 6 (January 1974): 41–47; Michael Brett, "Ibn Khaldun and the Invasion of Ifriqiya by the Banu Hilal (5th Century AH/11th Century AD)," in *Actes Du Colloque International Sur Ibn Khaldoun, Alger 21–26 Juin 1978* (Algiers: Société nationale d'édition et de diffusion, 1982), 289–98; Abdelmajid Hannoum, "Translation and the Colonial Imaginary: Ibn Khaldūn Orientalist," *History and Theory* 42, no. 1 (February 2003): 61–81; Allen Fromherz, *Ibn Khaldun: Life and Times* (Edinburgh: Edinburgh University Press, 2010); Robert Irwin, *Ibn Khaldun: An Intellectual Biography* (Princeton: Princeton University Press, 2018).

**85** The Zirid–Norman relationship is considered in David Abulafia, "The Norman Kingdom of Africa and the Norman Expeditions to Majorca and the Muslim Mediterranean," *Anglo-Norman Studies* 7 (1985): 26–49; Michael Brett, "Muslim Justice under Infidel Rule: The Normans in Ifriqiya,

Al-Idrisi wrote his description of Ifriqiya in the *Nuzhat al-Mushtaq* when the Normans controlled Mahdia and nearby cities in Ifriqiya. The decades-long involvement of the Normans in the region—both in times of peace and war—meant that al-Idrisi had access to numerous individuals with firsthand knowledge of Ifriqiya. Al-Idrisi also had a handful of earlier geographies to draw upon and use as a point of comparison for his work.[86] For example, al-Bakri (d. 1094) wrote about the wondrous ruins of the theater of Carthage, which contained columns adorned with engravings of animals and artisans. Al-Idrisi similarly writes that the columns of this theater contain depictions of people, artisans, animals, and boats. Many other elements of these descriptions vary (including the size and number of these columns as well as the history of Carthage more generally), but similarities like this indicate al-Idrisi's uncited use of this earlier source.

Although al-Idrisi used earlier sources in compiling his geography, he also added numerous contributions based on his own firsthand knowledge and information from interviews. The resulting description of Ifriqiya is extraordinary for its level of detail. Al-Idrisi records some 216 place names across this chapter (both human settlements and geographical features), the majority of which he places on the coastline. A central theme running through this description is one of economic and population decline brought about by both the Banu Hilal and, to a lesser extent, the Normans. The former of these groups is called the "Arabs" in this chapter, and their negative effect on the region is present throughout it—especially in inland areas. The markets and suburbs of Baghai, for example, "are now empty because of the pillaging of the Arabs." The city of Bilqan (near Gafsa) is similarly "now in ruins; Arabs defiled the city and took control of its benefits as well as all its lands and water." Outside the Mahdian suburb of Zawila, there "used to be a fortified area before the arrival of the Arabs into the land of Ifriqiya," whereupon they "desolated this land with its gardens, orchards, other wondrous fruits, and great goods." Al-Idrisi even describes an entire road from Gabès to Tripoli before noting that all of the places mentioned between these two cities are "uninhabited and deserted" due to attacks from the Arabs—so much so that "these lands have no amiable inhabitants nor allied residents."[87] The changes (namely the abandonment of settlements) caused by this political upheaval is such that many of the toponyms provided in this chapter have locations that we cannot reliably pinpoint on a modern map.[88]

---

517–55H/1123–1160AD," *Cahiers de Tunisie* 43 (1995): 325–68; Charles Dalli, "Bridging Europe and Africa: Norman Sicily's Other Kingdom," in *Bridging the Gaps: Sources, Methodology and Approaches to Religion in History*, ed. Joaquim Carvalho (Pisa: Pisa University Press, 2008), 77–93.

**86** Al-Bakri, *Description de l'Afrique septentrionale par el-Bekri*, trans. William McGuckin de Slane (Paris: Imprimerie Impériale, 1859); Al-Bakri, *Al-Masalik wa-l-Mamalik*, 2 vols. (Beirut: Dar al-Kotob al-Ilmiyah, 2003).

**87** See p. 62.

**88** There are many sites in Ifriqiya where we made educated guesses, many of them informed by the work of Bresc and Nef, about the modern ruins that correspond to them. This is an imprecise,

The political upheaval caused by the Banu Hilal in the 1050s is felt throughout al-Idrisi's description of Ifriqiya. His constant mention of these attacks shows their lasting effects on the people of the region, who suffered from their relentless incursions. We should, though, take these cataclysmic descriptions of the Banu Hilal with more than a grain of salt, as we know that some migrants to Sicily during this period were from ruling families displaced by the Banu Hilal and their successor dynasties—and who would have been all too eager to blame the present state of Ifriqiya on these outside invaders.[89] The utility of these passages on Arab devastation goes beyond the historical narrative they espouse, as these descriptions prove that al-Idrisi was not relying solely on older geographies, whose works do not consider the Banu Hilal. Instead, we can be confident that when al-Idrisi borrowed information from older geographies, he was not copying them mindlessly, but instead putting them in dialogue with the numerous contemporary sources to which he had access (in keeping with the methodologies outlined in the preface).

Al-Idrisi's critique of the Normans is more muted, which is understandable given his position at the royal court in Palermo. For example, al-Idrisi provides laudatory descriptions of major urban areas along the coast under Norman control, but also acknowledges that these places had seen better days. At Sfax, for example, he mentions bountiful fish harvests and markets of the city. He then notes that "the city was conquered by the great King Roger in the year 543H. Now, the city is still populous, but not as it once was with regard to its buildings, markets, and commerce." In his description of Mahdia, al-Idrisi similarly notes that "before the great King Roger conquered the city in the year 543H, Mahdia had vaults of gold that used to make its kings proud," which implies that now, the city does not have these same pride-worthy vaults of gold. At Annaba, too, al-Idrisi notes that the city "was conquered by the hand of one of the men of the great King Roger in the year 548H. Now, the city is weak and lacking in prosperity." Although al-Idrisi stops short of directly blaming the Normans for present decline in Ifriqiyan cities, the juxtaposition of Norman conquest with present downturn is evocative, and likely indicates that al-Idrisi had negative thoughts about these Ifriqiyan campaigns.[90]

The description of Ifriqiya provided by al-Idrisi thus emphasizes contemporary decline in Ifriqiya. The types of settlements that he describes in the region, especially on the coastline, further emphasize this fragmentation. Of the 92 inhabited settlements that he lists on the coast, 71 of them are fortresses or fortified villages (*qaṣr*, pl. *quṣūr*). The descriptions provided for many of these sites are very vague—sometimes only including a mention of a market or the strength of the walls—and do not show the same level

difficult, and (at times) very speculative process. Al-Idrisi, *La Première géographie de l'Occident*, 179–209.

**89** See Talbi, "Ibn S̲h̲addād," *EI2*.

**90** See p. 58. Other decisions within the *Nuzhat al-Mushtaq* perhaps reinforce this claim. Even though the Norman kings experimented with the title "King of Africa," al-Idrisi does not provide Roger II with this title. Similarly, the preface does not explicitly mention Norman conquests against the Zirids, though it does broadly mention Norman conquests against enemies to the east and west.

of concern as major urban areas along the coast. While al-Idrisi is careful to note which cities on the coast are presently held by the Normans, most of the coastal settlements in Ifriqiya have no noted owner or are recognized as parts of larger, vaguely defined parcels of land controlled by "Berbers" or "Arabs." This interpretation of Ifriqiya in the *Nuzhat al-Mushtaq* therefore emphasizes historical decline, whether due to Arab or Norman invasion, and the amorphous borders in Ifriqiyan settlements not under the latter's control.

The translation provided in this book includes the entirety of the second section of al-Idrisi's Third Clime, which is the section devoted to Ifriqiya, and a few short excerpts from the first section of the Third Clime (which considers the Maghreb to the west). This admittedly awkward division is due to the geographical limits of the Normans' conquests on the southern shores of the Mediterranean. The Normans launched attacks on the coastal cities of Barashk and Jijel during the course of their North African campaigns, which are located further west than the rest of their conquests spanning from Tripoli to Tunis. Al-Idrisi placed Barashk and Jijel as part of the eastern end of the Maghreb—and thus as part of the end of the previous section. In order to provide translations of the entirety of the North African lands under Norman control, we opted to include these brief passages in this book—though we acknowledge that it would be preferable to understand these entries alongside the numerous other entries for the first section of al-Idrisi's Third Clime (and the rest of his geography entirely).

## Within the *Nuzhat al-Mushtaq*: Sicily and Nearby Islands

Clime 4, Part 2 of the *Nuzhat al-Mushtaq* includes a description of the islands of the Tyrrhenian Sea (namely Sardinia and Corsica), the Aeolian islands, and the Norman-held island of Malta. The clear majority of this section, though, comprises al-Idrisi's description of Sicily. This description is one of the longest sections within the *Nuzhat al-Mushtaq*—rivaling those on the Maghreb, al-Andalus, and the Islamic heartlands in the Middle East—and indicates the importance of the island to al-Idrisi and, by extension, King Roger II. Indeed, the Norman court was based in Sicily, Roger owned a substantial amount of demesne land on the island, and trade routes passing through and within the island were a considerable source of income for the Norman administration. It was imperative that Roger and his advisors have an accurate understanding of the island, its people, infrastructure, and economic production. Al-Idrisi provides an unprecedented level of detail on toponymy, distances, fortifications, agriculture, commerce, natural resources, and topography of both coastal and inland areas of the island. In doing so, he elevates the status of Sicily as a place of both commercial productivity and defensive stability.

Al-Idrisi's description of Sicily surpasses in detail earlier attempts at describing the landscape of the island, including the work of the Roman-era geographer Strabo and the tenth-century Muslim scholar al-Maqdisi.[91] Muslim geographers of the

---

**91** Al-Maqdisi, for example, lists thirty settlements, all on the coast. He does, though, mention a few settlements that al-Idrisi does not, including the larger scale settlements of Rometta and Partanna.

Middle Ages, who tended to live and write in the Islamic heartlands of the eastern Mediterranean, tended to focus on Sicily's trade-facilitating coastal settlements. The *Nuzhat al-Mushtaq*'s Sicily section departs from this formula and is divided into three distinct categories—coastal settlements, interior settlements, and ports/anchorages of the island—that comprise around 150 place names in addition to more than 100 ports/anchorages.[92] Across this section, al-Idrisi provides details on a substantial number of interior settlements in Sicily, which could then be used by the Norman bureaucracy to determine how to best govern this sprawling island. Although it is difficult to plot many of these interior locations on modern maps, the mere mention of these settlements and the relative locations to one another make this description of Sicily an invaluable source for medieval Sicily.

Unlike his description of Ifriqiya, al-Idrisi's entry on Sicily has a fairly standard nomenclature for describing places of human settlement: cities (*madīna*, pl. *mudun*); defensive structures like forts (*qalʿa*, pl. *qalāʿa*) and castles (*ḥiṣn*, pl. *ḥuṣūn*); and rural settlements (*raḥl*, pl. *arḥul* and *manzil*, pl. *manāzil*). Of these settlements, cities are the least common but the most important. Al-Idrisi describes nine of them on the island (eight on the coast and one in the interior) with fittingly praiseworthy language. He considers both their historic roots and modern roles as crucial facilitators of commerce. Perhaps unsurprisingly, the city of Palermo—home to the court of Roger II—stands out among these cities for its grandeur:

> First is the city of Palermo, which is the great and beautiful city, the pulpit which stands above all the countries of the world. It is the proudest and most noble city and has been the seat of kings from the present to bygone days. It is the point from which troops and armies have set off to invade and the point to which they have returned, just as they still do nowadays. Palermo is situated on the coast, with the sea to the east, and great, towering mountains encircle it. The city's coast sparkles with cheerful radiance. The city boasts magnificent buildings which travellers to the city wander around wondering at their excellence, their detailed features, and the originality of their design.[93]

The laudatory descriptions of Sicilian cities within the *Nuzhat al-Mushtaq* does not extend to the many forts and castles mentioned therein. These structures, often built atop rocky hills, were built or fortified by the Normans so that they could retain control over the rugged interior of the island, especially along its northern half.[94] Al-Idrisi tends to only give a brief overview of these fortifications with a particular eye toward their defensive characteristics, freshwater sources, and commercial engagement.

---

al-Maqdisi, *Aḥsan al-taqāsīm fī maʿrifat al-aqālīm* (Cairo: Maktaba Madbouly, 1991); al-Maqdisi, *The Best Divisions for Knowledge of the Regions: Ahsan al-Taqasim Fi Marifat al-Aqalim*, trans. Basil Anthony Collins (Reading: Garnet Publishing, 2000).

**92** In Arabic, these categories are *al-bilād al-baḥrīyya*, *al-bilād al-barīyya*, and *mirās al-jazīra*.

**93** See p. 78.

**94** Nef, *Conquérir et gouverner la Sicile islamique aux XIe et XIIe siècles*, 394.

Take, for example, the fortress of *Qalʿat al-Nisà* ("Fort of the Women," which is modern Caltanissetta):

> Caltanissetta is a well-built fortress overlooking a continuous series of cultivated fields. It has excellent facilities, crops, trees and fruits and close by in an eastward direction flows the Salso River.

The utility for such descriptions is clear. Norman administrators reading (or hearing) this narrative would be presented with a basic description of the defensive capabilities of a fortress, a sense of how the people living there obtained their water (useful in the event of a siege), and what could be taxed in those lands. The final major category of human settlements that al-Idrisi considers in Sicily is the *raḥl* or *manzil*: unfortified rural settlements most similar to the "estate" or "manor" found in much of Europe. These settlements typically did not have walls or fortifications and were more akin to a land or fiscal unit—without a civic centre and associated with a benefactor.[95] Most of these unfortified settlements considered in the *Nuzhat al-Mushtaq* are located in Sicily's interior and comprise short descriptions of basic economic and geographic information (similar to forts and castles). The description of the estate of *Shalyata* is typical of these entries:

> *Shalyāṭa* is an estate on a plain with gushing rivers, fertile farmland, plentiful bounties and numerous crops. The *Al-ʿAsal* river flows in the western part of *Shalyāṭa's* territory and between *Shalyāṭa* and Piazza Armerina [*Iblāṭsa*] in a northerly direction is ten miles. The *al-ʿAsil* River emanates from Piazza Armerina.

The importance of trans-Mediterranean shipping to the Sicilian economy meant that knowledge of ports and anchorages across the island was imperative.[96] Al-Idrisi lists some 112 reliable places to drop anchor—spaced at an average of six miles apart. Some of these anchorages are well-established ports, such as Palermo and Syracuse, but many of them are smaller locations (often coves or inlets) where sailors could seek shelter from unfriendly winds and waves. The preponderance of Arabic place names for these locations indicates that many of them were frequented by sailors who spoke that language and/or were coined when Muslim lords governed Sicily. The utility of these descriptions, even if they are a simple listing of X port located however many miles from Y port, is clear. They allowed government administrators to better

---

**95** Ferdinando Maurici, *Castelli medievali in Sicilia: Dai bizantini ai normanni* (Palermo: Sellerio editore, 1992), 87, 119. Nef argues that the term *raḥl* in Sicily indicates an administrative and financial unit created by Norman lords. Nef, *Conquérir et gouverner la Sicile islamique aux XIe et XIIe siècles*, 403–4. See also Lucia Arcifa, "Alle origini del Raḥal di 'Abbûd: I documenti di età Normanna," in *L'insediamento rupestre di Monte S. Antonio a Regalbuto alle origini del Raḥal di 'Abbûd*, ed. Ileana Contino and Francesca Buscemi (Caltanissetta: Paruzzo Editore, 2012), 83–94.

**96** David Abulafia argues that the Normans were not "economically dynamic" in that they did not encourage manufacturing—except for the building of ships—but instead they preferred to export raw materials and tax goods passing through Sicilian ports. Abulafia, "The Crown and the Economy under Roger II and His Successors," 1–2.

inform sailors (whether transporting goods and/or people) of how to safely navigate around the island and where they could reliably drop anchor in the event of inhospitable weather.

Although the *Nuzhat al-Mushtaq*'s description of Sicily is unsurpassed for its analysis of settlements, it provides less details on the people of the island. Al-Idrisi only mentions a handful of local customs and history, like a place in Palermo where a Muslim sultan held court, but these accounts are few and far between. He only mentions a handful of religious sites, including a few mosques and churches, and does not consider in any detail the cultural diversity of the island, which historians know comprised (among other groups) Lombards, Greek Christians, Jews, Muslims, Arabs, and Berbers. The sole exception to this rule is the Normans themselves—led by Roger II—who feature prominently when al-Idrisi first introduces the island of Sicily. He describes the bounties of Sicily, the victorious conquests of Roger I, and the righteous expansion of Norman territories under Roger II before launching into a description of the island's settlements themselves. This description, which is reminiscent of the glowing language found in the preface, stands in stark contrast to al-Idrisi's lukewarm writings about Norman involvement in Ifriqiya as well as the general lack of a Norman presence in south Italy (to be considered shortly), which provides some indication of the importance of Sicily as the geographical crown jewel of the Norman monarchy.

Clime 4, Part 2 does not focus exclusively on Sicily, however. At the beginning of this chapter, al-Idrisi considers a handful of other islands in the central Mediterranean, though the amount of space afforded to each varies. Al-Idrisi provides the most detailed description of the island of Sardinia, in which he considers the basic geographical contours of the island and the three major settlements therein before turning to the people that inhabit it:

> The people of Sardinia are of Latin African [*Rum Afārqa*] origin, the most barbaric and savage of the races of the Rum; they are brave and determined and they do not relinquish their weapons. On the island there is a good silver mine and from this mine silver is exported to many countries of the Rum.[97]

Unlike most of the description of Sicily, in which al-Idrisi provides precious few details about the people and cultures of the island, this passage on Sardinia paints a very different picture of "Latin African" peoples with barbaric tendencies.[98] Perhaps a reflection of Norman anxiety at a possible alliance between the Pisans and Sardinians, this entry does not have the same lofty tone as those that describe Sicily.[99] Instead, this

---

**97** See p. 74.

**98** On medieval Sardinia, see Stephen L. Dyson and Robert J. Rowland, Jr., *Archaeology and History in Sardinia from the Stone Age to the Middle Ages: Shepherds, Sailors & Conquerors* (Philadelphia: University of Pennsylvania Press, 2007); Alex Metcalfe, Hervin Fernández-Aceves, and Marco Muresu, *The Making of Medieval Sardinia* (Leiden: Brill, 2021).

**99** There is substantial evidence for connections between North Africa and Sardinia up to the Muslim conquest of North Africa in the seventh century, see Raimondo Zucca, "Le Civitates Barbariae e l'occupazione militare della Sardegna: aspetti e confronti con l'Africa," in *L'Africa*

description is more reminiscent of the sections in which al-Idrisi describes non-Norman lands (like Ifriqiya) by describing basic geographical information about a location and any additional societal information that might have utility to the Norman court. Here, the mention of barbaric fighters and profitable silver mines conjures the image of a hostile people whose trade with potential rivals of the Normans could lead to problems in the future.

Other islands found in this chapter of the *Nuzhat al-Mushtaq* follow roughly this same formula, but in less detail. Al-Idrisi typically provides a brief geographical description of the shape of an island, its location relative to other sites, and a small piece of cultural and/or economic information that might be of utility to the Norman court. This is even true of islands that (apparently) were under Norman control, such as Pantelleria, Gozo, and Malta—which the Normans took during the course of their Ifriqiyan conquests in the 1120s.[100] Although al-Idrisi mentions these islands by name, he provides little information about them besides their dimensions, agricultural productivity, and distance to other islands. He does not mention the name of any settlements on the islands, as he did for Sardinia. The lack of detail that al-Idrisi provides for these Norman possessions is perplexing, especially when compared with the lofty description of Sicily that follows it. Perhaps knowledge about these lands was lacking, or the Norman court was less interested in them, or al-Idrisi struggled to find people with detailed knowledge about them, or al-Idrisi wanted to downplay Norman governance on these previously Muslim islands? In any case, the *Nuzhat al-Mushtaq* presents a very different kind of description of the Norman's non-Sicilian territories than its possessions on their main island.

## Within the *Nuzhat al-Mushtaq*: South Italy and the Adriatic

Norman holdings in south Italy and the eastern Adriatic do not fit neatly into a single section of the *Nuzhat al-Mushtaq*, but rather span a handful of chapters. Clime 4, Part 3 describes the coast of Calabria and Basilicata, southern Apulia, and a portion of the eastern Adriatic coast including the islands of Corfu and Cephalonia. Clime 5, Part 2 deals with the Tyrrenhian coast including the Norman mainland capital of Salerno,

*romana: atti del 5. Convegno di studio, 11–13 dicembre 1987, Sassari (Italia)*, ed. Leopoldo Ortu (Sassari: Università degli studi di Sassari, Dipartimento di Storia, 1988), 349–73; Manlio Brigaglia, Attilio Mastino, and Gian Giacomo Ortu, *Storia della Sardegna, 1, dalla preistoria all'età bizantina* (Rome: Editori Laterza, 2002); Leopoldo Ortu, *Storia della Sardegna: dal medioevo all'età contemporanea* (Cagliari: CUEC Editrice, 2011). Linguistic evidence further suggests a connection between Sardinian and African Latin. J. M. Adams, *The Regional Diversification of Latin 200 BC–AD 600* (New York: Cambridge University Press, 2007), 576.

**100** The Normans first conquered Malta during the reign of Roger I around 1090, though it is unclear if this was a lasting settlement or a raid on the island. Charles Dalli, "The Siculo-African Peace and Roger I's Annexation of Malta in 1091," in *De Triremibus: Festschrift in Honour of Joseph Muscat*, ed. Toni Cortis and Timothy Gambin (San Gwann: Publishers Enterprises, 2005), 265–74; Charles Dalli, "Satellite, Sentinel, Stepping Stone: Medieval Malta in Sicily's Orbit," in *Malta in the Hybleans, the Hybleans in Malta. Malta Negli Iblei, Gli Iblei a Malta*, ed. Anthony Bonanno and Pietro Militello (Palermo: Officina di studi medievali, 2008), 245–58.

Naples and Amalfi as well as the inland regions of Campania, including Benevento and Avellino. And Clime 5, Part 3 considers Norman territories in south Italy including the eastern Adriatic coast and inland Calabria, Basilicata, Apulia, Molise and Abruzzo. Across these chapters, al-Idrisi does not heap effusive praise upon Norman lands (as he does for Sicily), but instead provides relatively sparse details about the settlements found therein.

Although al-Idrisi details the distance between settlements in south Italy, he does not elaborate much on the settlements themselves. The details below on the settlements between Trani and Andria are representative of other entries in this chapter:

> Trani is a medium-sized city with a wall and a well-known market. From Trani to Barletta is six miles along the coast. Connecting with Barletta there is a city called Andria, situated on a plain, which is large and well-populated and between there and the sea is nine miles.

Within this description, we can see the same types of information present in the Sicily chapter, though without the same degree of pomp and circumstance. Al-Idrisi mentions basic information about a settlement (defensive capabilities, agricultural productivity, source of water, infrastructure, distance between other settlements) before moving onto the next location. The chapters on south Italy do not contain lofty panegyrics about Norman governance, but rather jump into the analysis of settlements (similar to his format in the Ifriqiya chapter). This is not to say, though, that al-Idrisi was in some way negligent in his analysis of the inhabited places of south Italy. Across the sections translated for this monograph, he compiles around one hundred settlements, more than any geography to date—especially in the context of Arabic ones, which had little to say about Christian-controlled south Italy.

The quantity of settlements considered within al-Idrisi's description of south Italy, when combined with the minimal details provided for them, might indicate the origins of the data that informed this section: the Norman military. Roger II waged campaigns across south Italy during the late 1120s and 1130s.[101] Within the court of Palermo, al-Idrisi likely had access to individuals involved in these expeditions who could provide him with details about the rugged landscape of the region. In the context of the Abruzzo region (east of Rome), Laurent Feller argues convincingly that "what we are told appears to have been drawn from the report of military reconnaissance...the text could thus serve as a guide for the staff officers charged with organising troop movements on the frontier, either for defence or for an offensive based upon fixed lines of communication."[102] It is possible too, following the argumentation of Jean-Charles Ducène, that al-Idrisi combined this information from military campaigns with other

**101** On Norman governance in south Italy, see Paul Oldfield, "Urban Government in South Italy ca. 1085–ca. 1127," *The English Historical Review* 122, no. 497 (2007): 579–608; Joanna Drell and Paul Oldfield, eds., *Rethinking Norman Italy: Studies in Honour of Graham A. Loud* (Manchester: Manchester University Press, 2021).

**102** Laurent Feller, "The Northern Frontier of Norman Italy, 1060–1140," in *The Society of Norman Italy*, ed. Graham Loud and Alex Metcalfe (Leiden: Brill, 2002), 66–67.

chronicles like that of Falco of Benevento, which considers Norman campaigns in south Italy.[103] Accounts from soldiers, rather than merchants or agricultural workers, would account for the lack of detail on agriculture and commerce—especially in comparison to the *Nuzhat al-Mushtaq*'s more substantial consideration of Sicilian settlements.[104]

Of particular importance to medieval armies was access to clean water and marching routes along major waterways. This significance is seen across the sections of the *Nuzhat al-Mushtaq* devoted to south Italy, wherein al-Idrisi pays particular attention to water sources. For example, the city of Otranto, which was a major Norman port and point of travel for individuals (both soldiers and pilgrims) going to the Holy Land, receives more detail than others around it:

> The city of Otranto is a city of ancient monuments and is very populous, and there are prosperous markets and lively commerce. The sea surrounds the city walls on three sides and the city is connected with the inland on its northern side. There is a river which descends from the north, then flows past the city gate and empties into the Adriatic Sea at Brindisi. There are forty miles between Brindisi and Otranto. The inhabitants of Brindisi are Lombards and previously the city was under the control of the lord of Constantinople. The city of Otranto is situated on the western side of the strait between the Mediterranean Sea and the Adriatic. From Otranto to the city of Durrës is seventy miles.[105]

This description of Otranto is found in Clime 4, Part 3 of the *Nuzhat al-Mushtaq*. A few other passages in Clime 5, Part 3 consider additional ways in which Otranto is connected to nearby coastal locations like Lecce and Brindisi. The utility of these entries for soldiers and merchants is apparent. Al-Idrisi provides them with an outline of what sources of water flow where and how close settlements along those waterways are to one another. This is particularly true of the Norman stronghold of Salerno, which was an important city during the Normans' conquests of south Italy in the 1130s. The description of it provided by al-Idrisi focuses on the water that surrounds the city and how to navigate around it:

> Salerno is an impressive city with plentiful markets, public utilities, wheat and grain. From Salerno to the Asa River, which is a narrow anchorage, is six miles and from there to the Sele is twelve miles. The Sele River has abundant water and is navigable, its banks are forested and swampy, rendering them difficult to access, therefore boats and fleets can dock there in safety. From there to the

**103** Jean-Charles Ducène, "Routes in Southern Italy in the Geographical Works of Al-Idrīsī," in *Journeying along Medieval Routes in Europe and the Middle East*, ed. Alison Gascoigne, Leonie Hicks, and Marianne O'Doherty (Turnhout: Brepols, 2016), 143–66.

**104** It is likely, too, that al-Idrisi used earlier Arabic geographies at times in this section. He, for example, echoes the categorization of Ibn Hawqal when he refers to Naples as a "city of linen." David Romney Smith, "Calamity and Transition: Re-Imagining Italian Trade in the Eleventh-Century Mediterranean," *Past & Present* 228, no. 1 (2015): 15–56.

**105** See pp. 105–6.

> gulf of Agropoli and the island of Licosa close to the mainland is twenty miles, and there is no port there.[106]

The soldiers that likely informed al-Idrisi of the geography of south Italy and its nearby islands also told the geographer of defensive fortifications at given locations. We can see this trend when al-Idrisi describes the islands of the eastern Mediterranean where the Normans had long fought against Byzantine forces.[107] When describing the island of Corfu, for example, al-Idrisi reports the following:

> And west and slightly to the south of this island is the island of Corfu. This is a large island; its length is one hundred miles. There is a city which is well-populated, fertile and well defended and the city is situated on an impregnable pinnacle. The city's people are ready to defend themselves against those that attack them.[108]

Perhaps referencing longstanding conflict between the Normans and Byzantines in the region, al-Idrisi notes that the people of Corfu are willing to defend themselves against attackers. Indeed, it is possible that al-Idrisi was conducting research on this island as Norman forces were trying to conquer it, seeing as George of Antioch captured the island in 1147 and held it until 1149.[109]

With these sections on south Italy, although al-Idrisi rarely launches into discussions of the people and history of given locations, he does make clear the lands held by the "Franks" (Normans) and the "Lombards," whom the Normans often fought. In Clime 5, Part 3, he describes how numerous settlements on the eastern coast of the Adriatic, including Bari, Brindisi, Conversano, Campomarino and Matera are "part of the countries of the Lombards."[110] Given that these territories were by this time under Norman control, albeit in league with local lords, al-Idrisi could be here speaking historically, in reference to the duchies of Spoleto and Benevento, the latter of which had absorbed Byzantine Apulia.[111] Al-Idrisi's distinction between Lombard and Frank is in keeping with practices elsewhere in south Italy, as Graham Loud has argued that charters up to the 1150s were clear in stipulating whether someone was Norman or Lombard.[112] Joanna H. Drell has more recently argued that there was a subtle resis-

**106** See pp. 109–11.

**107** William B. McQueen, "Relations between the Normans and Byzantium 1071–1112," *Byzantion* 56 (1986): 427–76; Paul Stephenson, *Byzantium's Balkan Frontier: A Political Study of the Northern Balkans* (New York: Cambridge University Press, 2000), 223–24.

**108** See p. 106.

**109** Stanton, *Norman Naval Operations in the Mediterranean*, 94–101.

**110** See p. 113.

**111** Errico Cuozzo, "Intorno alla prima contea normanna nell'Italia meridionale," in *Cavalieri alla conquista del Sud: Studi sull'Italia normanna in memoria di Lèon-Robert Ménager*, ed. Errico Cuozzo and Jean-Marie Martin (Rome: Laterza, 1998), 171–93.

**112** Graham Loud, "How 'Norman' Was the Norman Conquest of Southern Italy?," *Nottingham Medieval Studies* 25 (1981): 13–34.

tance to cultural absorption by both Norman settlers and local Lombards and that it was thanks to King Roger's "tolerance of the variety of local customs in his kingdom that there were no successful rebellions during his reign and the people of the South continued to identify themselves as Norman or Lombard, sometimes even Greek until the end of the twelfth century."[113] While many towns may be in Lombard hands in Apulia, according to al-Idrisi, the settlements of what he terms "the country of Calabria" including Catanzaro, Viggiano, Benevento, Melfi, Chiaromonte and Senise are all under Frankish control. In contrast to the close association between the Norman monarch and Sicily, the Normans have a much less clear presence in south Italy except when al-Idrisi specifies that these lands belong to the "Franks."[114]

## Our Translation

This monograph is the result of collaboration between four scholars with different backgrounds and areas of academic expertise, and we divided our efforts based on our respective strengths. Katherine Jacka was the primary translator for the sections on Sicily and south Italy, Sherif Abdelkarim provided the translation for the preface, and Matt King wrote the translation for Ifriqiya. Ahmad Kzzo provided extensive editorial feedback for all of these translations, while all four authors gave comments to the others on their respective sections. Katherine and Matt were the primary authors of this historical introduction. Katherine was also the driving force in pushing this manuscript to publication and corresponding with the editors at Arc Humanities Press.

The authors used a handful of dictionaries for this translation. *Lisan al-Arab*, an Arabic–Arabic dictionary finished in 1290 by Ibn Manzur, was the most useful near-contemporary work for the *Nuzhat al-Mushtaq*.[115] The *Hans Wehr Dictionary of Modern Written Arabic* (fourth edition) provided further clarification on direct English translations of al-Idrisi's Arabic.[116] *Lane's Lexicon* and *al-Faraid Arabic-English Dictionary* were likewise used as necessary.[117] These dictionaries, combined with frequent references to the French and Italian translations considered above alongside the *Encyclopaedia of Islam* (both second and third editions) were essential in puzzling together al-Idrisi's prose.

**113** Joanna Drell, "Cultural Syncretism and Ethnic Identity: The Norman 'Conquest' of South Italy and Sicily," *Journal of Medieval History* 25 (1999): 202.

**114** The use of the term "Franks" is further compelling because it is commonly used in Arabic geographies to describe lands under Christian control. Perhaps al-Idrisi was transposing this language upon these lands, even though much of the information he received about them came from Norman (or "Frankish") soldiers?

**115** Ibn Manzur, *Lisan al-ʿArab*, 3rd ed., 20 vols. (Dar Sadir, 2000).

**116** Hans Wehr, *Arabic–English Dictionary: The Hans Wehr Dictionary of Modern Written Arabic*, ed. J. M. Cowan, 4th ed. (Ithaca: Spoken Language Services, 1994).

**117** Edward Lane, *Arabic–English Lexicon*, 8 vols. (London: Williams & Norgate, 1863); J. G. Hava, *Al-Faraid Arabic–English Dictionary* (Beirut: Dar al-Mashriq, 1970).

Across this monograph, we have endeavored to standardize our translations in keeping with the original Arabic of al-Idrisi. This has been a difficult task. The act of translation is inherently one of linguistic domestication that (to some degree) flattens and distorts the original text. This is particularly true of Arabic-to-English translation, which is difficult enough in itself—and is even tougher when the Arabic was written in the Middle Ages. Al-Idrisi's language (especially in the preface) is often incredibly hard to render into modern English, and at times we had to make educated guesses about the intent behind his words. We have tried to be transparent about these issues, which we describe in footnotes accompanying our translation. Despite the obstacles inherent in this kind of translation, we have sought to standardize our language for certain terms that occur throughout the work of al-Idrisi ("*madīna*" is translated as "city," for example) so that readers have some sense of the vocabulary used across sections. For the ease of modern readers, we have also organized our translation into paragraphs of digestible length that are united by common themes and ideas.

We have also tried to make the Arabic of al-Idrisi accessible to modern audiences. When al-Idrisi first mentions a settlement that corresponds to a modern toponym that resembles the original Arabic, we have provided that modern toponym. Thus, we introduce the city of Gabès without the accompanying Arabic since the transliteration (*Qābis*) is reminiscent of the modern name. If a modern toponym departs substantially from the Arabic that al-Idrisi gives, then we have indicated the Arabic transliteration of that location. Thus, when al-Idrisi first introduces us to the site of *Mūnsa*, which corresponds to modern Palmarola in Italy, we provide the following: Palmarola (*Mūnsa*). We hope that this system makes the toponomy of al-Idrisi accessible to modern readers while reducing the clutter that comes with frequent references to the original Arabic. When we do use transliterations, we adhere to the guidelines provided by the *International Journal of Middle East Studies*. In general, these transliterations avoid diacritics except for technical terms and prefer English-language translations when available. Our one intentional deviation from these guidelines is for settlement names, for which we have provided diacritics so that those with knowledge of Arabic can reverse engineer the spelling of any given location. To avoid any confusion over the names of settlements, we have also provided an index that includes the modern place name (if known), original Arabic and transliterated Arabic for all locations considered in this monograph.

*KITAB NUZHAT AL-MUSHTAQ FI IKHTIRAQ AL-AFAQ*

# THE BOOK OF PLEASURE FOR THOSE WHO DESIRE TO VISIT FARAWAY PLACES

## TRANSLATION

# PREFACE OF THE *NUZHAT AL-MUSHTAQ*

BY THE NAME of Allah, Possessor of Mercy, Dispenser of Mercy. Bless, o Allah, our leader Muhammad, his family. Protect them.

Praise Allah! Owner of the might, the glory; the largesse, appreciation; the benefit, bestowal, great favours; who planned then executed, provided then prospered; ordained then upheld, prepared then perfected; created and innovated, honed what he fashioned. Thus did minds come to know him: arose in souls his authority; enlightened for eyes his proof; defied intellects his power, prerogative. Guide to the path of his glory by grace and direction; designator of his blessings with himself by speech and doctrine; maker of his strange creatures and amazing inventions an avenue toward knowing him, a ladder toward knowing his ancientness, his eternity. There surely inheres in what he created a lesson for possessors of vision, a reminder for those with sharp intellects and thoughts. Among his signs are the creation of the heavens and earth. Concerning the sky, he raised its ceiling and sorted its transit; adorned it with stars and lodged in it the sun and the moon—two signs, sought to illumine night and day and know by their circuit the sequence of eras and epochs. As for the earth, he levelled its ground, erected its mountains, extracted its water and pasture, and therein housed his creation. And so, he bestowed them its dominions: tamed for them its stars; informed them of its courses; taught them its benefits and harms. He guided them to travel through it—by land and sea smooth and rugged. All that derives from him, his power radiating precision, arrangement, will, decision. Exalted, from this, his reign, authority, craft, and proof.

Now, the best to have utilized his vision, applying it toward unprecedented thoughts and considerations, must be the exalted King Roger, empowered by Allah the self-powered, king of Sicily, Italy, Lombardy, and Calabria; the victorious Roman leader of the Nazarene faith, in that he excels the Byzantine king in the distribution, acquisition, and delivery of matters, ratifying and voiding as he pleases. He dealt among his faithful justly, protecting them under the wing of grace and favour. He raised with the ropes of his rule an excellent effectuation: he handled the flow of his polity according to the best organization, the most beautiful society, and opened lands east and west. He subdued the necks of the tyrants among his fold far and near with what he possesses from well-equipped troops of ready numbers, and heavy fleets—the concerted reinforcement. They proved true in experience and intelligence; their hearing and sight matched in acuity.

What distant objective, then, failed to reach him or escaped his purview? What weighty ambition did he not attain or meet with ease his reach? For the fates flow according to his desires and designs; good fortunes serve him, disposing his choices' every movement and halt. His friends abide in adamant, eminent honour; his enemies, unceasing disgrace and ruin. How many stations of honour did he bolster? How many endeavours' merits did he extract? He brought out their moons and lit their terrains; their gardens became lush meadows, their crops prosperous. Moreover, he added to fine manners noble ancestry, and to beautiful deeds an excellent disposition, plus spiritual courage; mental purity; intellectual depth; abundant understanding; sound judgement and planning; expertise in the execution of affairs out of a complete, piercing perception. His pursuits are like the striking arrow; barred businesses unlock in his hands, while all politics depend on him. His sleep equals others' wakefulness, his decrees the fairest decrees. His gifts: copious seas, delivering cloudbursts.

Regarding his expertise in the fields of mathematics and the applied sciences, it cannot be enumerated, much less specified, because he took from their every art a massive share: he's like a brimming vessel, boiling over. He contrived some unusual inventions and devised some strange devices—what no kings have previously done, nor did any outmatch him—and this here constitutes evidence for witnesses, clear confirmation and proof: their travels through the cities and regions have obviated our mentioning them distinctly and diversely. Doing so would divide rather than gather matters, although were we to describe them and reflect on their composition and assembly, his signs would amaze us, their meanings being miraculous, their objectives dear. Yet who would count the number of pebbles and reach thereby the ultimate end?

Among his splendid expertise, then, and lofty, noble tendencies: when the deeds of his kingdom expanded, and the zeal of his nation's people increased, and the Roman countries obeyed him, its peoples coming under his loyalty and authority, he wished to know the properties of his lands truly; to study them definitively and expertly; to know their limits and methods by land and sea, be they in any Clime; and what distinguished them as to the seas and gulfs situated about them, along with knowledge of other lands and regions within the seven Climes, upon which the theologians agreed and the records of the copyists and authors verified; and to which of each Clime the lands cleaved and referred and belonged, according to the written books on this subject. Among the science of all these is the example of al-Masʿudi's *The Book of Wonders*; the book of Abu Nasr Saʿid al-Jihani; the book of Abu al-Qasim ʿUbayd Allah ibn Khardadhabah; the book of Ahmad ibn ʿUmar al-ʿUdhri; the book of Abu al-Qasim Muhammad al-Hawqali al-Baghdadi; the book of Khanakh ibn Khaqan al-Kimaki; the book of Musa ibn Qasim al-Qaridi; the book of Ahmad ibn Yaʿqub, known as al-Yaʿqubi; the book of Ishaq ibn al-Hasan al-Munajjim; the book of Qudamah al-Basri; the book of Claudius Ptolemy; and the book of Arsius of Antioch.[1]

Yet he did not find therein a detailed, comprehensive annotation. Rather, he found these sources inane. So he summoned the experts of this matter, consulting and conversing with them about it. He then realized that they possessed no more knowledge than what was in the aforementioned books. When he saw them in this state, he called across his lands. The experts then attended in response, those esteemed in these studies. He asked them accordingly, collectively and individually, yet they arrived at no consensus. He approved the collection of their most resolved and enduring reports; whatever they disagreed upon he set aside or discarded. He maintained this procedure for about fifteen years, never depriving himself at any time from examining this field—its inquiry into it and research into its truth—until he attained what he desired from it. Moreover, he desired to ascertain with certainty the soundness of the consensus his band of counsellors arrived at concerning the account of the altitudes and

1 These sources are considered in al-Idrisi, *La Première géographie de l'Occident*, 27–30.

widths of the lands. Therefore, a drawing board was brought to him; he set to testing it with iron measurements little by little while referring to the aforementioned books and weighing the sayings of its authors.

He scrutinized all this until he arrived at the truth therein. Thereupon, he ordered that a purely silver sphere be made available to him, comprehensive and massive: bulky, weighing four hundred Roman rottols.[2] Each rottol costs one hundred and twelve dirhams. When it was completed, he ordered the labourers to engrave upon it pictures of the seven Climes with their lands; their shoreside and countryside; their gulfs, seas, and channels; their rivers' positions; its arable and barren areas; what lies between each of their lands and the others in terms of trodden paths and specified miles and attested distances and known ports, according to what comes to them—[all this] represented in a drawing map. They left nothing out of it, and produced it in its shape and look, as prescribed. And [he ordered them] to author a book that conforms to its forms and appearances, except that it add to the former by describing the conditions of the lands and terrains in their nature, location, positions; their shores, seas, mountains, distances; their crops, expenses, kinds of buildings; their properties; the tools they use; the industries that pay; the businesses that attract and export; the wonders said of and ascribed to them; where they fall within the seven Climes, with mention of the conditions of their people: their bearings, manners, orientations, fashions, clothing, and languages; and to name this book: *The Book of Pleasure for those who Desire to Visit Faraway Places*. This was done in the first ten days of January, agreeing with the month of Shawwal in the year 548.[3] And so, I herein heed the injunction and articulate the drawing.

Regarding this discussion, I first begin on the subject of the shape of the earth, known as geography, as Ptolemy named and described it. From Allah we seek aid, success, and fulfilment in every method and course. He—glorious his power—is fit for this, proficient. And so, we say that what can be gleaned from the discussions of the philosophers, the leading scholars, and the experts in astronomy, is that the earth is round as a ball. The water cleaves to it, resting atop it in natural inertia, not leaving it. The earth and the water are fixed within the sphere like yoke in the belly of the egg. Their atmosphere is a moderate one. The wind surrounds them from all directions; it attracts them to the side of the sphere, or propels them. Allah knows best about the truth of this.

The earth is fixed within the sphere. This is so on account of the severity of the sphere's speed. All the creatures lie on its surface. The wind attracts the light matter in their bodies, and the earth attracts the heavy matter in their bodies by the power of the magnetic rock that pulls iron to it. The earth is divided into two parts, between which is the equator, from the east to the west. This is the length of the earth; it is the largest line on the globe, just as the zodiac is the largest line in the celestial sphere. The earth's circumference at the equator is three hundred and sixty degrees: the degree is twenty-five farsangs; the farsang is twelve thousand cubits; the cubit is

---

**2** From the Arabic *raṭl*, a historical measure of dry weight equivalent to one to five pounds.

**3** January 1154.

twenty-four fingers; the finger is six grain seeds, arranged in a contiguous line, anterior touching posterior.[4] With this ascription, the earth's encirclement becomes one hundred thousand thousand cubits, thirty-two thousand thousand cubits. In farsangs, it becomes eleven thousand farsangs—this according to the calculation of the people of India. As for Hermes, he measured the circumference of the earth and made every part one hundred miles; by this [measurement], it becomes thirty-six thousand miles, and becomes, in farsangs, twelve thousand farsangs. Between the equator and the two poles are ninety degrees, likewise its roundness in breadth.[5]

Except for the life on earth beyond the equator that is sixty-four degrees, the rest of the earth is empty, devoid of life due to the extreme cold and frozenness. In general, all creation exists on the northern quadrant of the earth. Accordingly, the southern quadrant above the equator is neither populated nor inhabited due to the intensity of its heat, always in this state until its waters desiccated and its animals and plants died out from lack of humidity, since no animal or plant can ever exist except where there are waters and moistures. The earth itself is round, though not truly so: parts of it are depressed and elevated. The water runs in it from its highest to lowest point, and the surrounding sea encircles half the earth in an unbroken enclosure; its circle resembles a sphere, of which only half appears, as though, upon description, it were an egg submerged in water, with the water in a basin. Likewise, then, is the earth: half of it is sunk in the sea. The atmosphere surrounds the sea, and the atmosphere pushes or pulls them both, as we stated before this.

The scientists divided this inhabited fourth of the earth into seven Climes, each of which passes the equator from the west to the east. These Climes do not lie along natural lines, but rather certain imaginary lines, found by means of astronomy. In each Clime, there are a number of cities, citadels, towns, and nations; none of them resembles the others. Moreover, in every Clime there exist high mountains; contiguous lowlands; running springs and rivers; stagnant ponds; and various metals, plants, and animals. We will relate most of this in what's to follow, by Allah's favour and help.

Seven seas pierce these seven Climes; they are called gulfs. Six of them are connected, and one is separated; it does not connect with anything from the aforementioned seas. One of these seas on the inhabited earth is the Sea of China, India, Sindh, and Yemen. Its source from the east lies above the equator by thirteen degrees, stretched with the equator toward the west. It then passes China firstly, then India, then Sindh, then Yemen to its south, terminating at the Gate of Wailing [*al-Mandab*].[6] There, the sum of its height and the height of these distances, according to what the trusted travellers and seafarers who investigate it have reported (the sailors from land to land, from the inlet of the Red Sea [*Baḥr al-Qulzum*][7] to the *Wāq Wāq*), is four thou-

**4** One *farsakh* equals approximately six kilometres or just under four miles.

**5** Eratosthenes of Cyrene (d. 194 BCE) was the first person to calculate the circumference of the earth.

**6** This is the strait between Yemen and the Horn of Africa.

**7** This is literally "The Sea of Calm."

sand, five hundred farsangs.[8] It has about three hundred islands, inhabited and uninhabited. We will mention some of them afterward, if we've ascertained knowledge of it and the reports about it prove correct.

From this Chinese Sea branches the Green Gulf, which is the Sea of Persia, and *al-Ubulla*.[9] Its passage from the south to the north inclines slightly to the west, and so it passes the western lands of Sindh, Makran,[10] Kerman,[11] and Persia, until it stops at *al-Ubulla* where Abadan lies. There, the last of it ends, then its coast bends, returning to the direction of the south. It passes the lands of Bahrain and *al-Yamama*,[12] and connects with Oman and *al-Shihr*[13] from the lands of Yemen. Therein lies its connection with the Chinese Sea. The height of this sea equals four hundred and forty farsangs. It has the mountains of *ʿUwayr* and *Kusayr*.[14] Its depth is seventy to eighty fathoms. Among its islands are nine islands, inhabited and uninhabited. We will mention them in due course, with the aid of exalted Allah.

From this Chinese Sea branches also the Gulf of Suez [*al-Qulzum*]. Its source is from the Gate of Wailing and where the Indian Sea ends. Then it passes in the northern direction, slightly westward, and then connects with western Yemen. It passes through Tihamah[15] and the Hijaz to Midian,[16] Ayla,[17] and Paran,[18] until it ends at the

---

**8** The existence of the fabled *Wāq Wāq* island to the far east is disputed. Sufyan Yasen Ibrahim, "Exploration Of The Geographical Location Of Waq Waq Islands In The Light of Islamic Arab Geographical Sources," *Alustath Journal for Human and Social Science* 2, no. 225 (2018): 51–64.

**9** *Al-Ubulla* was a town of mediaeval Iraq situated at the head of the Persian Gulf and famed as the terminal for commerce from India and further east, see "al-Ubulla," *EI2*.

**10** Makran, the coastal region of Balochistan, was first invaded by the forces of ʿUmar bin Khattab in 634 CE. André Wink, *Al-Hind: The Making of the Indo-Islamic World*, vol. 1 (Leiden: Brill, 1991), 129–34.

**11** Kerman is ancient Carmania—another of the areas conquered by ʿUmar's forces in 634. Xavier de Planhol and Bernard Hourcade, "KERMAN Ii. Historical Geography," in *Encyclopædia Iranica*, n.d., www.iranicaonline.org/articles/kerman-historical-geography, accessed November 21, 2023.

**12** In early Islamic times, *al-Yamāma* was a region of Najd (in modern day Saudi Arabia). See "al-Yamāma," *EI2*.

**13** Now in Yemen, throughout the centuries the town has been associated with the incense trade. Ibn Kh̲urradadh̲bih calls the area the Land of Incense (*bilād al-kundur*). See "al-S̲h̲iḥr," *EI2*.

**14** These locations were mentioned by al-Masʿudi in his description of the islands of the al-Salamah archipelago at the northernmost point of Oman. John Craven Wilkinson, "A Sketch of the Historical Geography of the Trucial Oman down to the Beginning of the Sixteenth Century," *The Geographical Journal* 130 (1964): 337–49.

**15** Red sea coastal plain stretching from Bab al-Mandab to the Gulf of Aqaba.

**16** Midian was an ancient region located in northwestern Arabia. Compared with other peoples of the ancient Near East, knowledge about Midian and the Midianites is limited and restricted to a few written sources, particularly the Hebrew Bible and the Quran. Juan Manuel Tebes, "Midian," in *Oxford Bibliographies*, 2020, www.oxfordbibliographies.com/display/document/obo-9780195393361/obo-9780195393361-0271.xml, accessed November 21, 2023.

**17** The ruins of ancient Ayla which are found in central Aqaba. "Ayla," *EI2*.

**18** Some Islamic polemicists argue that Paran, found in Deuteronomy 33:2, refers to Muhammed

city of Qulzum,[19] to which it is ascribed. Its coast then bends, returning in the southern direction. It passes the eastern lands of Upper Egypt [*al-Ṣaʿīd*] to the Bay of the King [*al-Malik*];[20] then it proceeds from Aydhab to the Peninsula of Suakin,[21] to Zeila, from the lands of the Beja,[22] and ends at the lands of Abyssinia [*al-Ḥabashah*], connecting to the Indian Sea. The length of this sea is one thousand, four hundred miles. The deepest part of this sea consists of abysses on which ships crash. Therefore, only the masters sail it, the knowledgeable of its abysses, the adept in its courses and streams. Among its islands are fifteen islands. We will mention them with meticulousness at their proper point of their mentioning with Allah's aid.

As for the second great sea, known as the Mediterranean [*al-Baḥr al-Shāmī*],[23] its offshoot is from the Atlantic [*al-Baḥr al- Muẓlum*][24] that lies in the western direction, and its starting point is in the fourth Clime. There, it is called the Strait of Gibraltar [*al-Baḥr al-Zuqāq*][25] because its width there becomes eighteen miles. Likewise, the length of the passageway from the Island of Tarifa to Algeciras [*al-Khaḍrāʾ*][26] measures eighteen miles. It then passes eastward in the direction of the Berber lands, and northern Morocco, until it passes the central Maghreb and connects to the territory of Ifriqiya; to the *al-Wādi al-Raml*;[27] to Barqah,[28] Libya, and *Marāqiyah*;[29] to Alexandria; to the north of the territory of *Tīh*;[30] to Palestine and all the lands of Syria; until its tip reaches *al-Suwaydiyyah*,[31] its end-point. From there, its coast bends back, westwardly passing the lands of Antioch. It connects with the peninsula of Constantinople to the

and Mecca. Hagga Mazuz, "Tracing Possible Jewish Influence on a Common Islamic Commentary on Deuteronomium 33: 2," *The Journal of Jewish Studies* 67 (2016): 291–304.

**19** This is ancient Clysma.

**20** This location is uncertain.

**21** By the twelfth century, ʿAydhab had become the main entrepôt through which Indian spices were imported to Egypt. Suakin was a staging post en route. Andrew Peacock, "Suakin: A Northeast African Port in the Ottoman Empire," *Northeast African Studies* 12, no. 1 (2012): 29–50.

**22** This is a reference to the Beja people of the coastal areas of the Eastern Desert.

**23** On this etymology, see "al-Shām," *EI2*.

**24** This means "The Dark Sea" which applied to the north Atlantic as descriptive of its bad weather and dangerous character. See "al-Baḥr al-Muḥīṭ," *EI2*.

**25** This is literally "The Corridor Sea."

**26** This is literally "The Green Island."

**27** This is literally "The Valley of Sand," the location of which is uncertain.

**28** An ancient city, Barqah came to be used in the Islamic period as the name for the whole region of what had been Cyrenaica.

**29** This is an ancient town west of Alexandria.

**30** This refers to the desert forming the frontier zone between Palestine and Egypt and spanning the Sinai Desert. See "al-Tīh," *EI2*.

**31** According to the geographer Yaqut al-Hamawi, *al-Suwaydāʾ* was founded by the Nabataeans and lies en route to Sham by two nights' travel. Yaqut al-Hamawi, *Muʿjam Al-Buldan*, vol. 3 (Beirut: Dar Sadir, 1995), 286.

peninsula of Peloponnese [*Bulbūnus*] and to Otranto. The Adriatic [*Jūn al-Banādiqa*][32] commences there and connects to the Strait of Sicily; the lands of the Rum;[33] the lands of Savona and Narbonne. It courses the Pyrenees [*Jabal al-Burtāt*], then passes the eastern lands of *al-Andalus* toward its south-central, and ends at the two peninsulas whence it began. The length of the Mediterranean from its beginning to end is one thousand, one hundred and thirty-six farsangs. Among the islands within it are about one hundred islands, small and large, inhabited and uninhabited. We will mention them when the place of mentioning arrives, with Allah's aid.

Two gulfs proceed from this Mediterranean. One of them is the Adriatic; its initial point starts east of the lands of Calabria from the lands of the Rum, from the city of Otranto, passing in the northern direction with a slight western bent. It then passes the territory of Bari to the shore of Sant'Angelo. It then runs west to the lands of Ancona until it passes the coast of Venice and concludes at the lands of Aquileia. From there, its coast bends back, returning eastwardly to the lands of Croatia [*Jarūsiyya*], Dalmatia, and the lands of Slavonia [*Isqlūniyya*] until it connects with the Mediterranean, where it began. The length of this peninsula from its initial point to its very end is one thousand, one hundred miles. Among its islands are fifteen, from them six inhabited ones. The rest are uninhabited. We will mention this in its proper place.

From the Mediterranean emerges also a second peninsula called the Black Sea [*Nīṭs*].[34] It begins from the Mediterranean where the mouth of the Dardanelles lies. The width of its mouth there is an arrow's throw. It passes two routes then connects with Constantinople. There, its width becomes four miles, and it likewise passes sixty miles until it reaches the Black Sea in the eastern direction. It then connects from the south with the territory of Heraclea to the territory of Sinop [*Istrūblī*]; to the coasts of Trabzon; to the territory of *Ashkālah*;[35] to the territory of the Alans [*Lāniyah*].[36] The edge of this peninsula ends there, where the Khazars live.[37] From there, its coast bends, returning to *Maṭrakha*; it connects with the Russian lands, the lands of Bulgaria [*Barjān*], and the site of the Dnieper [*Danābars*] river.[38] It passes to the Danube [*Dunu*]

32 This is literally, "The Gulf of Venice."

33 The term "Rum" (usually translated as "Roman") has a handful of potential meanings in the work of al-Idrisi that is dependent on the context. Here al-Idrisi is likely referring to the Papal States. At other times, though, the term is used in a chronological context to refer to either the ancient Romans or the Byzantines. "Rūm," *EI2*.

34 The geographer Muḥammad al-Himyari (d. 1495) locates the sea of Nīṭs in the Black Sea.

35 This location is uncertain.

36 The Alans were an ancient Iranian tribe of the northern (Scythian, Saka, Sarmatian, Massagete) group, known to classical writers from the first centuries CE. V.I. Abaev and H.W. Bailey, "ALANS," in *Encyclopædia Iranica*, 1985, https://iranicaonline.org/articles/alans-an-ancient-iranian-tribe-of-the-northern-scythian-saka-sarmatian-massagete-group-known-to-classical-writers-from, accessed November 21, 2023.

37 The Khazars were a nomadic people in the South Russian steppes who flourished in the early Islamic period. See "Khazar," *EI2*.

38 *Maṭrakhah* is ancient Phanagoria.

river until it ends at the mouth of the Constantinople Peninsula. It connects with Constantinople and passes the eastern lands of Macedonia, until it connects to the place where it began. The length of the Sea of *Nīṭs*, from the mouth of the passage to its end-point is one thousand, three hundred miles. In it are six islands. We will mention them upon arriving at the point of their mentioning, with the aid of exalted Allah.

As for the Sea of Gorgan[39] and the Daylam,[40] it is a cut-off sea: it does not connect with any of the aforementioned seas. Several rivers and ever-flowing springs fall within it. From the east, the territory of the Turks [*al-Aghzāz*] connects with this sea; from the north, the territory of the Khazar; from the west, the lands of Azerbaijan and the Daylam; and from the south, the lands of Tabaristan.[41] Its length from Khazar to the Spring of Distress [*al-Hamm*][42] is one thousand miles. Its width from the direction of Georgia [*Jarjān*] to the site of the Volga [*Ithl*] river is six hundred and fifty miles. Within it are four islands; their description will come later.

On each sea from all the seas that we brought up are lands and nations that we will mention with explanation and clarification, land by land, nation by nation, with exalted Allah's aid. Within these seas additionally lie various fish and animals, of various characteristics; the wonderful ones we will describe in their proper place of mentioning with exalted Allah's aid.

Having finished mentioning the shape of the earth—its division of Climes and mentioning of seas that we mentioned in its physical properties, extents, boundaries, and what follows its coasts in terms of lands, nations, and all that concisely—let us now begin mentioning the seven Climes and what they encompass of lands, nations, and wonders, Clime by Clime, land by land. We mention what they include in terms of kingdoms, and will provide their roads, paths, and the sum of their farsangs, miles, riverways; the height of their seas, the ways of their deserts. All this, explicated and clarified in a concise, descriptive inquiry, with the utmost effort and maximum energy exerted. With Allah lies success. From him comes aid. To him is the power, the might.

When we wished to draw these cities in the Climes (and their paths, and what encompasses their nations), we divided the length of each Clime into ten sections: parts measured by length and width. We drew in each section what belonged to it from cities, villages, and social centres in order for the onlooker to see in this what has been hidden from his eyes, or what he did not know, or what he could not reach on account of the difficulty of the roads and the dissimilarity of the nations, and so the information matches the eye-witness account. The sum total of the following drawings after this is seventy pictures, not including the final two, the first of which is the final habitation in the south, most of it empty due to the extreme heat and lack of water;

**39** The Gorgon Gulf is at the southern end of the Caspian Sea, though al-Idrisi is talking about the Caspian Sea as a whole.

**40** Daylam was a shifting region in northern Persia and adjacent territories, including the Deylaman uplands.

**41** This is the name for the southern shore of the Caspian Sea (in northern Persia). See "Ṭabaristān," *EI2*.

**42** This location is uncertain.

and the second inhabited end in the north, most of which empty due to the extreme cold. It has also become clear—with what we have mentioned and detailed—that the viewer, should he look on to these visual descriptions and aforementioned lands, will have seen from them an accurate positioning and pleasant semblance. However, it remains on him after this to know the qualities of the kingdoms and conditions of the nations: their adornments, dress, and travelled roads, with their miles and farsangs; and the wonders of their lands from what the travellers witnessed, what the roamers reported, and what the transmitters verified. For this reason, we have decided to mention after every image what deserves commentary in accordance with its placement in the book, based on knowledge and competence, and Allah is the helper. No other master exists.

# CLIME THREE, PART ONE (Partial)

### Pages 257–258

BARASHK IS A small city on a hill and surrounded by an earthen wall.[1] It is on the coast. Its people drink from springs, and its water is fresh. The great King Roger conquered Barashk in the year 5__.[2] Barashk has fruits and various farmlands, which yield much wheat and barley. From there to Cherchell is twenty miles. Between Barashk and Cherchell is an impenetrable mountain, where a tribe from the Berbers called the Rabiʿa live.[3]

### Page 268

Jijel is a city on the coast, which surrounds it, and it has a suburb. When the fleet of the great King Roger triumphed there, the people of Jijel fled the city to a mountain that is around a mile away, where they constructed a fortified city. During the winter, they lived in the port and the coast. During the summer (meaning the time of the departure of the fleet), they moved their belongings and the entirety of their goods to this highest castle far from the sea.[4] The men, though, remained on the coast with a few of their goods for trading. Now, Jijel is in ruins; its houses and walls are destroyed, and there are no inhabitants there or near there. Jijel used to be a charming city with milk,

---

**1** The modern location of Barashk is uncertain. Agabi advocates that it was located on the ancient site of Gunugus, while Richards (following the argumentation of Forstner) sees it as the modern location of Sidi Brahim (ancient Bencenna) in modern-day Algeria. This location is recorded in the works of Ibn Hawqal and Leo Africanus, though there is no record of inhabitation of Barashk after the sixteenth century. C. Agabi, "Breshk/Barashk," in *Encyclopédie Berbère*, 1992, https://doi.org/10.4000/encyclopedieberbere 1870 [accessed November 21, 2023]; Ibn al-Athir, *The Chronicle of Ibn Al-Athir for the Crusading Period from al-Kāmil Fī'l-Ta'rīkh*, trans. D. S. Richards, vol. 1 (Aldershot: Ashgate, 2006), 375.

**2** Within the manuscripts of al-Idrisi, the last two years of this date are not legible. It is likely that Roger II conquered Barashk around 539H (1144–45) based on the testimony of Ibn al-Athir. Ibn al-Athir, *Al-Kamil fi al-Tarikh*, vol. 11 (Beirut: Dar Sadir, 2009), 51–53; Ibn al-Athir, *The Chronicle of Ibn Al-Athir for the Crusading Period from al-Kāmil Fī'l-Ta'rīkh*, 1:482–83.

**3** The Rabiʿa were an Ifriqiyan tribe that had lived in the region since at least the ninth century CE. Lorenzo M. Bondioli, "Islamic Bari between the Aghlabids and the Two Empires," in *The Aghlabids and Their Neighbors: Art and Material Culture in Ninth-Century North Africa*, ed. Glaire D. Anderson, Corisande Fenwick, and Mariam Rosser-Owen (Leiden: Brill, 2017), 482–83.

**4** The weather of the Mediterranean was such that it was safer to sail in the summer than in the winter, especially for military fleets. There are a handful of examples from Arabic chronicles of the Zirid dynasty, for example, losing its fleet in attempts to sail to Sicily during the winter. Thus, the people of Jijel were more comfortable living on the coastline during these winter months when local traders might have been able to frequent their ports, but not hostile ships coming from Norman Sicily. Abraham Udovitch, "Time, the Sea and Society: Duration of Commercial Voyages on the Southern Shores of the Mediterranean during the High Middle Ages," in *La Navigazione Mediterranea Nell'alto Medioevo* (Spoleto, 1978), 503–46; John Pryor, "Winds, Waves, and Rocks: Routes and the Perils Along Them," in *Maritime Aspects of Migration*, ed. Klaus Friedland (Köln: Böhlau Verlag, 1989), 71–85.

ghee, honey, and many crops. The fish there are of great numbers, infinite in goodness and quantity.

**Pages 268–269**

The city of Jijel also has two ports. The one on the south side is a crude port that is hard to enter without a skilled guide. Conversely, the port on the north side is called the port of the Woods [*Marsa al-Shaʿrāʾ*], for it is tranquil in movement like a good basin. It is good for moorage, but it cannot hold large ships because of its small size and sand.

# CLIME THREE, PART TWO

HERE BEGINS THE Second Part of the Third Clime, which contains cities, regions, castles, strongholds, peoples, and nations. Within these lands are Qamuda, Baghai, Meskiana, Majana, Béja, Annaba [*Būna*], El Kala [*Marsa al-Kharaz*], Bizerte, Lorbeus, Marmajana, Qastiliya, Bilqan, Degache [*Taqiyus*], Zarat [*Zarūd*], Gafsa, Nefta, El Hamma du Jérid [*al-Ḥamma*], Tunis, Kelibia, Hergla, Sousse, Mahdia, Sfax, Gabès, Raghugha, Mansouria [*Ṣabra*], Tripoli, and Lebda.[1] Along the coast of this sea in this chapter are castles, outposts, and cultivated lands that we will consider in what comes after this, with the help of Allah.

Baghai is a large city encircled by two walls made of stone. The suburb around it has a wall as well, and there used to be markets in it. However, the markets in the city and suburbs are now empty because of the pillaging of the Arabs. Baghai is located at the beginning of the lands that grow dates. A river runs to the city from the direction of the qibla.[2] Its people drink from it, and they also drink from fresh-water wells. The city used to have grazing lands, villages, and cultivated lands. Now, all of that is diminished. Around Baghai are the cultivated lands of the Berbers, who do business with the Arabs, and most of their crops are wheat and barley. The seizure of taxes and the administration of the city's affairs is undertaken by their shaykhs. Baghai connects to Mount Aurès at a distance of some miles.[3] Its length is around twelve days, and its people are overlords of those nearby. From the city of Baghai to Constantine is a three-day journey. From Baghai to Tobna (of the Zab) is a four-day journey.[4] From Baghai to the city of Qastiliya is a four-day journey.

Qastiliya is also called Tozeur. It has a fortified wall and very many date palms. Its many dates are common in the lands of Ifriqiya. There, the citron is big, good, and delicious. Most of the fruits there are grown in moderate quantities, and the legumes are plentiful and of high quality. Tozeur's water is not good and does not hydrate well. The cost of food there at most times is high because it is imported. The crops of wheat and barley in Tozeur are of a small quantity. Tozeur connects to the city of El Hamma du

---

**1** We will consider these locations as they occur within this chapter. It is worth noting, though, that al-Idrisi does not mention Qamuda in this chapter or anywhere else in his geography. This location, the specific location of which is unknown, was the primary town of the region of Gammouda in modern Tunisia. The works below are perhaps the most useful overall for mapping out the cities found in al-Idrisi. Ernest Babelon, René Cagnat, and Salomon Reinach, *Atlas archéologique de la Tunisie* (Paris: Ministère de la Guerre, 1893); Stéphane Gsell, *Atlas archéologique de l'Algérie* (Algiers: A. Jourdan, 1911); Paul-Louis Cambuzat, *L'évolution des cites du Tell en Ifrîkiya du VII.° au XI.° siécle*, 2 vols. (Argel: Office des Publications Universitaires, 1986).

**2** The qibla faces the direction of the Kaaba in Mecca—in this case, east.

**3** Al-Idrisi is likely referring broadly to the Aurès Mountains, which are located at the eastern edge of the Atlas Mountains in modern Tunisia and Algeria.

**4** Tobna is located near the modern city of Barika in Algeria on the site of ancient Thubunae. The arrival of the Banu Hilal brought about the apparent decline of the city, as the description of the city in al-Bakri is more detailed and positive than that of al-Idrisi. "Ṭubna," *EI2*.

Jérid from the southeast. Between these two cities is a short day's journey. The water of El Hamma du Jérid is not good, but the drink satisfies its people. There, date palms are many and dates are abundant. From there to Degache is around twenty miles.[5] Degache is a charming city located between El Hamma du Jérid and Gafsa. Degache is a city full of henna, cumin, and caraway. Its date palms and dates are of excellent quality, and it has an abundance of good and soft legumes. From Degache to the city of Gafsa is a day's journey.

The city of Gafsa is a charming city with a wall and a flowing river with water better than that of Qastiliya. In the middle of the city is the spring called al-Tarmidh. Gafsa has prosperous markets, many shops, and various industries. The city is surrounded by many date palms, which contain a variety of wondrous dates. It also has beautiful gardens, orchards, as well as erected and inhabited forts scattered about the city. Varieties of crops—like henna, cotton, and kumin—are grown here. The people of Gafsa are barbarians, and many of them speak in the African Latin tongue.[6]

From the city of Gafsa to the southwest is connected the city of Qasara, which has been mentioned previously.[7] The city of Naqawas and the city of Jamunas are to the east of it.[8] All of these lands are similar in their situation and have declined in their qualities. Date palms, water, crops, and wheat are always little because the majority of them are imported. The city of Gafsa is the centre of the lands that encircle it. From Gafsa to the city of Kairouan to the northeast is a four-day journey. To the southwest is the city of Bilqan, which is a five-day journey.[9] Bilqan is now in ruins; Arabs defiled the city and took control of its benefits as well as all its lands and water. From there to Gafsa is a four-day journey. From Gafsa to the south (on the side of Mount Nafusa) is the city of Zarat.[10] Between the cities is a five-day journey. From the city of Gafsa to the

**5** Degache is located adjacent to the ancient city of Thiges, from which the Arabic *Taqiyus* originates. Ducène, *L'Afrique dans le Uns al-Muhağ Wa-Rawḍ al-Furağ d'al-Idrīsī*, 159.

**6** Scholars have written much about this passage, as it indicates the presence of Latin Christianity in inland Ifriqiya in the middle of the twelfth century. Although scholars have long posited that indigenous Christianity in North Africa had long been on the decline in the early and central Middle Ages, this brief passage testifies to the existence of some Christian communities that had (at least) weathered the invasions of the Banu Hilal. Al-Idrisi uses similar language to refer to the people of Sardinia, which could indicate a perceived connection between them. Henri Bresc, "Le royaume normand d'Afrique et l'archevêché de Mahdiyya," *Le partage du monde: Échanges et colonisation dans la Méditerranée médiévale*, 347–66; Dominique Valérian, "La permanence du christianisme au Maghreb: l'apport problématique des sources latines," in *Islamisation et arabisation de l'Occident musulman (VIIe–XIIe siècle)*, ed. Dominique Valérian (Paris: Éditions de la Sorbonne, 2011), 131–49; Heinz Halm, *The Empire of the Mahdi: The Rise of the Fatimids* (Leiden: Brill, 2021), 98–99.

**7** Despite what al-Idrisi writes, this is his first (and only) mention in this text of Qasara, the location of which is unknown.

**8** The locations of Naqawas and Jamunas are unknown and especially difficult to triangulate given the ambiguity of the settlement of Qasara.

**9** The location of Bilqan is unknown.

**10** Mount Nafusa is presumably the western part of the Nafusa Mountains, which stretch from central Tunisia through western Libya. The location of the settlement that al-Idrisi calls *Zarūd*, though, is uncertain. Bresc and Nef place it at modern Zarat, which is located southeast of Gabès

city of Nefta is a short two-day journey. Nefta is a civilized city full of people, markets, trade, date palms, crops, and running water. From Gafsa to Nefzaoua in the south is two days and part of a day.[11] From Tozeur to Nefzaoua is a day and half of a long day.

From the city of Gafsa to Mount Nafusa in the south is around six days. This mountain is tall; it is around three days in length or less. On it are two minbars for two cities.[12] One of these cities on the mountain is called Sharwas, which has running water, vineyards, good grapes, and figs.[13] The bulk of their crops is good barley of the utmost quality. When it is made into bread, it is the tastiest of all the food of other regions. Sharwas' people are skilled in making bread; their expertise is more honed than any other group. From the city of Gafsa to the city of Sfax is three days.

Between Mount Nafusa and the city of Nefzaoua is the city of Luhaqa, which connects in the west to the cities of Biskra and Badis.[14] All of these lands are similar to each other in their sizes and qualities, and in their shops and markets. From Mount Nafusa to Ouargla is a twelve-day journey. From Nefta to the city of Gabès is a journey of a little more than three days. Gabès is a splendid and prosperous city. It is surrounded on all sides by forests, intersecting gardens, gardens in lines, and cheap, general fruits. It also has dates, farms, and agricultural lands that are surrounded on the outside by a trench. The city has markets, buildings, trade, and goods. It used to have a workshop that produced good quality silk, but now it has tanneries for leather products, which are used as equipment. Gabès has a river that comes to it from a large stream.

Along this stream is the fort of Saja, which is a small and civilized city.[15] Between it and Gabès is three miles. Located on the coast, Saja has a market, vendors, and many silk sellers. Its people drink from the river of Gabès. The water of the city of Gabès is not good, but it is drinkable, and its people find it palatable. Between the city of Gabès

---

and near the edge of the Nafusa Mountains, and we have followed suit here. Al-Idrisi, *La Première géographie de l'Occident*, 181.

**11** The Nefzaoua (also rendered as "Nefzawa") were a Berber group that lived in Ifriqiya during the Middle Ages. Here, though, al-Idrisi refers to Nefzaoua as a city more so than a tribe. Based on the location of these lands in this geography, al-Idrisi is referring to the area (and possibly a population centre there like modern Kebili or Bechri) southeast of the Chott el Djerid, which is a large salt lake in west-central Tunisia.

**12** The minbar is the elevated platform (pulpit) from which the imam delivers a sermon in a mosque. In this case, al-Idrisi indicates the presence of a mosque in each of the two mentioned cities.

**13** Sharwas was a city in the Nafusa Mountains (western Libya near modern Buqayqilah) that was important enough to be mentioned in the works of Ibn al-Warraq, Ibn Hawqal, al-Bakri, and al-Idrisi. A handful of excavations at the site indicate that it was abandoned sometime in the early twelfth century. Olwen Brogan, "Annual Report of the Society for Libyan Studies," *Libyan Studies* 2 (1971): 11; N. M. Lowick, "The Arabic Inscriptions on the Mosque of Abū Macrūf at Sharwas," *Libyan Studies* 5 (1973): 14–19.

**14** The locations of Luhaqa, Biskra, and Badis are unclear, though they might correspond roughly to settlements near modern Kebili—perhaps Bchelli, Bazma, or Bechini would make sense for the latter two.

**15** The location of this fort is unknown, though it is likely part of a suburb of Gabès along the Mediterranean coast.

and the sea is six miles to the north. Between the edge of the city's forest and the sea is a mile-long stretch of sand. This forest has trees, gardens, vineyards, and plentiful olives from which the people of the city produce oil, which is then exported to other locations. At Gabès, there are intertwined date palms full of fresh dates, which do not deviate at all from the pinnacle of taste. Likewise, the people of Gabès harvest these soft dates, then place them into earthen jugs. After a while, honey appears on the outside of the dates, and one is not able to eat the dates until it is removed. There are no dates in all of the lands famous for dates that match them, imitate them, or compare to their chewiness and good taste.

The port of Gabès on the sea is inconsequential because is not shielded from the wind. Instead, boats anchor at its river, which is small and affected by the tide of the sea. Small boats anchor within it, but not big ones due to their size. The tide of the sea rises for docking at a distance of around the length of an arrow shot. The people of Gabès lack refinement, but they have fine clothes and cleanliness. In its countryside, though, there is aggression, corruption, and highway robbers.

From the city of Gabès to the city of Sfax (following the coast of the bay) is seventy miles. Between the city of Sfax and Gafsa to the southwest is three days. Sfax is a long-populated city with many markets and substantial prosperity. It is surrounded by a stone wall that has gates with impenetrable iron plates. On its walls are admirably constructed guard towers for protecting the ribat.[16] Its markets are busy, and its people drink from cisterns. Unique and valuable types of fruits are imported to Sfax from the city of Gabès. They come in plentiful quantities sufficient for the city's needs and cheap in price. Fish, which are dangerous to catch and thus high in price, are caught in Sfax. Most are caught with pre-set traps, which are put in still water with crafty skill. Olives and oil, which are not found like this in any other place, are not comparable to others similar to them. Sfax has a good port with still water. Overall, it is a city with power in these lands; its people are proud and have honour in their souls. The city was conquered by the great King Roger in the year 543H.[17] Now, the city is still populous, but not as it once was with regard to its prosperity, markets, and commerce.

It is a two-day journey from Sfax to the city of Mahdia, where a governor in the service of the great King Roger lives. Mahdia is a city that continues to serve as a hub for the ships of traders coming and going from the lands of the Mashriq, Maghreb, al-Andalus, the lands of the Rum, and other places.[18] Many goods used to be brought to the city with huge sums of money over the years. However, trade has diminished

**16** The ribat was a defensive structure erected by Muslim armies as frontier outposts that could house soldiers and protect commercial networks from hostile forces. They were particularly common in North Africa as Muslim forces expanded their territories during the seventh and eighth centuries. "Ribāṭ," *EI2*.

**17** Sfax was conquered in the summer of 1148 as part of a larger campaign that saw much of the coast of Ifriqiya seized by the Normans. King, *Dynasties Intertwined*, 103–34.

**18** We follow here the work of Bresc and Nef, who take the word *al-ḥijāziya* as *al-jiḥāziya* (rendered in their translation as "bateaux marchands." It is possible, though, that al-Idrisi is referring to traders from the Hijaz (i.e. Arabian Peninsula) exchanging their wares in Mahdia. Al-Idrisi, *La Première géographie de l'Occident*, 184.

in our time. The city of Mahdia was the anchorage and port for Kairouan. The Mahdi, ʿUbayd Allah, founded the city in his name.[19] It is located on the coast. To travel there from Sfax, go to Raqqada of Kairouan, then travel to Mahdia from the city of Raqqada. The city of Mahdia from the city of Kairouan is a two-day journey.

Travellers used to frequent Mahdia a great deal. Merchandise was brought to the city from other lands and regions. Markets and goods were sold there; its people were ambitious and pursued success. Mahdia has fine buildings that are charming and clean structures, its facilities and abodes are lovely, its bathhouses are sublime, and it has many places to stay. At its core, Mahdia is beautiful in appearance from the inside and outside. Its people are beautiful in appearance and the cleanliness of their clothing. Lovely, intricate, and well-made clothes are produced there, which are transported and brought by traders to all distant lands at all times. There are no clothes made with the same ability or wonderful quality in any other lands or regions.

The people of Mahdia drink from cisterns, as its wells do not have a fresh taste. A strong wall built of stone surrounds Mahdia. Along it are two gates made of and strengthened by iron sheets without wood. The two gates are a marvel of construction. There is not known to be anything like them in inhabited lands in terms of workmanship and strength—they are wonders to describe! Within Mahdia, there are no gardens or orchards of date palms. Instead, fruits are brought to the city from the forts of Monastir—and between Mahdia and Monastir by sea is thirty miles. Monastir consists of three forts where pious worshippers live, and the Arabs do not harm anything at all—from their trees to their cultivated lands. In this place (I mean Monastir), the people of Mahdia bury their dead, which they carry on boats and then bury there. Then they return to their land, for there is no cemetery to be found in Mahdia in our time.

At the time of writing this book, Mahdia comprises two cities, the first of which is the city of Mahdia and the second of which is the city of Zawila.[20] The city of Mahdia is where the sultan and his troops live. His fortress is an impressive building, wonderfully constructed and made with skill. Before the great King Roger conquered the city in the year 543H, Mahdia had vaults of gold that used to make its kings proud. When Roger conquered Mahdia, its sultan was named al-Hasan ibn ʿAli ibn Yahya ibn Tamim ibn al-Muʿizz ibn Badis ibn al-Mansur ibn Buluggin ibn Ziri al-Sanhaji.[21]

**19** The city of Mahdia was founded in the early tenth century by the messianic Mahdi of the Fatimid Dynasty, ʿUbayd Allah, who belonged to a sect of Shia Muslims known as Ismaʿilis. The construction of this city on the coast of Ifriqiya provided the Fatimids with a secure base from which they could secure their authority in the region (which was a hotbed of Maliki Sunni jurisprudence) and prepare for future expansion to the east. Indeed, by the end of the tenth century, the Fatimids had relocated to Egypt and were preparing to launch attacks against the Abbasid heartlands in the Middle East. Michael Brett, *The Rise of the Fatimids: The World of the Mediterranean and the Middle East in the Fourth Century of the Hijra, Tenth Century CE* (Leiden: Brill, 2001).

**20** Today, Zawila has been subsumed by the city of Mahdia. It should not be confused with the other city of Zawila, located in southern Libya.

**21** Roger II (via his admiral, George of Antioch) conquered Mahdia in the summer of 1148, which effectively brought an end to the Zirid dynasty. The Zirid emir, al-Hasan, fled with his family west

In the city of Zawila, there are beautiful markets, fine buildings, wide streets, and spacious alleys. Its people are wealthy merchants—nobles endowed with insightful minds and thoughtful ideas. Their white dress is sublime. For these people and their clothes, there is beauty. They have extensive knowledge of commerce and their conduct is laudable in its practice. Zawila has very high and impregnable walls that surround the city by land and sea. They are built of stone. It also has many funduqs and bathhouses.[22] From the side of land, Zawila has a large trench filled with rainwater. Outside the city, toward the west, there used to be a fortified area before the arrival of the Arabs into the land of Ifriqiya. However, they desolated this land with its gardens, orchards, other wondrous fruits, and great goods. Today, nothing remains of this aforementioned land. Near Zawila, there are many villages, houses, and forts where people live. They have bountiful crops and livestock (namely sheep and cattle). They also produce much wheat and barley, and they extract much olive oil from plentiful olives—a wondrous product that pervades other parts of Ifriqiya and is sold to other lands in the Mashriq. Between these two cities—meaning Mahdia and Zawila—is a large space called the Sand [*al-Ramla*] that spans more than a length the shot of an arrow. Mahdia is the capital of the lands of Ifriqiya and the hub of its kingdom.

Since we have finished discussing our recollection of the lands of Ifriqiya, we will now return to a recollection of the lands of Nefzaoua.[23] Thus, we say that the city of Sbeitla was the city of Jarjis—the king of the African Rum.[24] It had the best appearance in the land, the most territory, the most abundant water, the fairest air, and the finest soil. Sbeitla also had orchards and gardens. Muslims conquered the city in the early years of Islam and killed its great king called Jarjis. From there to the city of Gafsa is a little more than a day's journey. Also, from Sbeitla to Kairouan is seventy miles.

The city of Kairouan is the mother of metropolises and the capital of the region. It was the greatest of the cities of the western region. It had the most people, the most free-flowing money, the most expansive prosperity, the most masterful buildings, the most precious activities, the greatest profits for trade, the most income from taxes, the most spending on commodities, the most profits, but also the most disobedient and the most filled with hate. Generally, though, the virtues of Kairouan's people are noble: devotion to excellence, fulfilment of contracts, relinquishment of suspicions, avoidance of the forbidden, mastery in the beauties of science, and a drive toward purpose.

---

before being captured by his Hammadid rivals. Eventually, al-Hasan was released from prison and made a co-governor of Mahdia by the Almohads. King, *Dynasties Intertwined*, 166–97.

**22** Funduqs were accommodations for merchants travelling through a location. They became particularly common in mediaeval North Africa as segregated housing for Christian merchants doing business in Muslim lands. Olivia Remie Constable, *Housing the Stranger in the Mediterranean World: Lodging, Trade, and Travel in Late Antiquity and the Middle Ages* (New York: Cambridge University Press, 2003).

**23** See note 15 in this chapter.

**24** Al-Idrisi is referring to Gregory the Patrician, a Byzantine Exarch of Africa who was killed in combat in 647 near Sbeitla while fighting Muslim armies. Allen Fromherz, *The Near West: Medieval North Africa, Latin Europe and the Mediterranean in the Second Axial Age* (Edinburgh: Edinburgh University Press, 2016), 89–90.

But Allah—Glory to Him!—imposed the Arabs on Kairouan, and calamities enveloped the city until no one remained in it except its obsolete ruins and the vestiges of its erasure.

Now, in our time, Kairouan is a fraction of itself. It is surrounded by an earthen wall. Its governors are the Arabs, and they collect what is available in taxes. Few people are there, their trade is minimal, and its profits are trivial. What people of discernment predict is that the city will come back soon with its prosperity and other things as it was once before. The city's water is scarce, and its people drink the water from the large cistern that is there. This cistern is a remarkable structure because it was built in a square shape and in its middle is a standing structure like a silo. Each of its sides measures two hundred cubits, and the whole thing is filled with water.

Kairouan consisted of two cities: the first was Kairouan and the second was Mansouria, which was the home of the king. In the days of its inhabitation, it had three hundred baths, most of which were for royal residences, the rest of which were out in the open for all the people. In our time, Mansouria is now in ruins. Not a single inhabitant is there. Three miles from it, there are the forts of Raqqada with their tall spires, many well-built orchards, and crops. At the time of Aghlabids' state and its splendour, the population used to spend the springtime there.[25] Now, Mansouria is in ruins, does not await power, and will not return to prosperity.

From the city of Kairouan to the city of Tunis is a journey of a little more than two days by a walking caravan. Tunis is a charming city that is narrowly encircled on all sides by flat lands and farms of wheat and barley, which are most of the crops that its people trade with trust in the Arabs and their emirs. The city is now, during our writing of this book, populated and replete with blessings such that people from near and far seek sanctuary in it. Around Tunis is a firm earthen wall that has three gates. All the gardens and farms of legumes of the city are inside of this wall and there is nothing outside of the wall that can be relied upon. The Arabs live adjacent to this land, and they bring many types of grain—as well as honey and ghee—in abundance for its inhabitants. They also make many types of bread that are not possible to make in any other lands.

The city of Tunis itself is ancient, eternal, and fortified. Historically, its name was Tarshish, but when Muslims conquered the city and constructed buildings in it, they named it Tunis.[26] Its people drink from various wells, but the greatest of them in mea-

---

**25** The Aghlabids ruled over much of Ifriqiya during the ninth and early tenth century before being conquered by the Fatimids. Their capital from 876 onward was Raqqada, which is directly southwest of Kairouan. Mohamed Talbi, *L'émirat aghlabide (184/860—296/909)* (Paris: Adrien-Maisonneuve, 1966); Mounira Chapoutot-Remadi, "Comment les Aghlabides ont-ils gouverné l'Ifriqiya?," in *The Aghlabids and their Neighbors: Art and Material Culture in Ninth-Century North Africa*, ed. Glaire D. Anderson, Corisande Fenwick, and Mariam Rosser-Owen (Leiden: Brill, 2017), 49–75.

**26** Tarshish is a biblical term that refers to a location far away from the eastern Mediterranean. Scholars tend to identify the city with Tartessos (Iberia), Tarsus (Asia Minor), or Taros (Sardinia). Al-Idrisi, however, associates Tarshish with the city of Tunis. Goldenberg speculates that Tarshish might have been used as a broad, catch-all term for Black Africa in certain rabbinic and targumic

sure and the one with the sweetest of water are two wells dug by women of Islam, who were looking for a reward. They were at the end of their strength, when a huge amount of water appeared. Tunis is the neighbour of Carthage, which is famous for its goodness, its quantity of fruits, its good location, the quality of crops, and the breadth of its crops. Among the crops there are cotton, hemp, caraway, and madder.[27] In our time, though, Carthage is in ruins and without any inhabitants.

The city of Tunis is in the middle of a bay (located outside of the sea), and is on an excavated lake, the width of which is more than its length—for its length is six miles and its width is eight miles. It has a mouth that connects to the sea, which is called the Mouth of the River [*Fam al-Wadī*].[28] This lake previously did not exist; it was dug from land via excavations that ended at the city of Tunis, for between Tunis and the sea is six miles (as we described previously). The width of this excavated river is around forty cubits, its depth is three to four fathoms, and its bottom is clay. The length of this trench, which is called a river, is four miles. The water of the sea was dragged over that excavated river (really over the trench) until the water exceeded a fourth of a fathom (more or less). When the water reached its top, it stopped. At the end of this trench, the water widens and deepens. This area is called the Venerable [*Wuqūr*] and to it arrive transportation ships, attack ships, and scouting ships that anchor there. The overflow of this floating water flows through this excavated river up to the city of Tunis, which is at the throat of the lake. The freight of ships is unloaded at the Venerable, where small boats float to the waters just short of the city of Tunis. Boats enter the river from the sea one after another to reach the Venerable, because the capacity of the river is such that it cannot host more than this. Some of this lake is connected to the west such that between it and Carthage is two miles. From the mouth of this lake to the city of Carthage is three and a half miles.

Carthage is now in ruins. However, people live in it on an elevated parcel of land called al-Muʿallaqa, which is surrounded by an earthen wall and is inhabited by leaders from the Arabs known as the Banu Ziyad.[29] The city of Carthage was (at the time of its construction) among the most extraordinary of the known lands because of its wondrous buildings and the demonstration of the craftsmanship required to build them.[30]

---

literature. David M. Goldenberg, *The Curse of Ham: Race and Slavery in Early Judaism, Christianity, and Islam* (Princeton: Princeton University Press, 2009), 41–43.

**27** Radder was used for dyeing textiles, as its roots (when processed) yield an orange-red color. Daniel V. Thompson, *The Materials and Techniques of Medieval Paintings* (New York: Dover, 2012), 121–24.

**28** This area is called La Goulette in modern Tunis.

**29** On the Banu Ziyad and al-Muʿallaqa, see Matt King, "An Emir in the Ruins of Carthage: The Life and Times of Muhriz Ibn Ziyad (d. 1160 CE)," *Journal of North African Studies* 28 (2023): 230–57.

**30** These structures are considered in Giovanna Vitelli, *Islamic Carthage: The Archaeological, Historical, and Ceramic Evidence* (Carthage: Centre d'etudes et de documentation archéologique de Carthage, 1981); Susan T. Stevens, "Carthage in Transition: From Late Byzantine City to Medieval Villages," in *North Africa under Byzantium and Early Islam*, ed. Susan T. Stevens and Jonathan P. Conant (Washington, DC: Dumbarton Oaks, 2016), 89–104; Dexter Hoyos, *Carthage: A Biography* (New York: Taylor & Francis, 2020).

Now, it has remnants of the known buildings of the Rum such as the theatre, which has no equal among the buildings of the earth with respect to its craftsmanship and size. Indeed, this theatre has the structure of a circle, which is around fifty arches that soar into the air. The size of each arch is more than thirty spans. Between each arch and its sister is a magnificent column. The size of each column and their supports is four and a half spans. Standing upon each arch from these arches are five additional arches—one after the other—each with the same character and construction of porous stone that is not like any other kind with respect to quality. Atop each arch from this surface is a circular medallion, which depicts images, and there is a circular sea on the lower surface, on which is depicted a variety of images and engravings on wondrous statues standing fixed in stone—including people, artisans, animals, and boats. All of these carvings were made with masterful craftsmanship and the most pronounced wisdom. The upper part of this towering building is very smooth and does not have a thing on it. It is said that this building used to be a stadium and a gathering place during a given season and day of the year.

Among the wonders of construction at Carthage are the cisterns, of which there are twenty-four in a line. The length of each cistern is one hundred and thirty steps by twenty-six steps in width. Each cistern has a vault above it. Between each cistern and its neighbour are holes and passages that bring water from one to the other. All of them have been constructed with great wisdom. Water used to run to these cisterns from the spring of Ain Djoukar [*Shūqār*], which is close to Kairouan.[31] The distance that this water ran from the spring to the cistern was three days. The running of water from this spring to these cisterns via this aqueduct was on a number of arches—their number was uncountable!—and a moderate amount of water ran over them. These solid arches were built of stone: those that were on high land were short, while those that were on the belly and depth of the earth were very tall. It is one of the most marvellous things I have seen on the face of earth! In our time, water is cut off from these cisterns, and nothing connects to them. All of this happened because of the destruction of the city of Carthage. Nevertheless, from the day of its destruction to the present, excavations in the ruins of forts and the foundations of buildings have provided different kinds of marble—too many to describe. Indeed, a report from an expert on the subject mentioned that he saw slabs removed from the marble that were a length of forty spans and a width of seven spans (and others less). Excavations in its ruins are always happening without interruption. The removal of marble from there does not cease, and its marble is carried to all regions of the earth. One does not depart from there in a ship or by other means unless they carry with them large quantities of its marble such that they become famous for it. One may find there columns of marble that have an entire circumference of forty spans (or less).

The city of Carthage is surrounded by flat lands and plains, which have farms and varieties of crops with many benefits. The land of Carthage is connected from the west

**31** This structure (in modern parlance) is called the Zaghouan Aqueduct.

with the region of the city of Satfura.[32] This is a splendid region that has three cities. The closest of them to Tunis are Ashluna, Tinja, and Bizerte. This latter city on the sea is fortified and smaller than the city of Sousse in size. Between Tunis and Bizerte is a long day of travel over land. The city of Bizerte is small but full of people. It has facilities and recurring markets for itself. On its eastern side is a well-known lake that is attributed to it. This lake is sixteen miles in length, eight miles in width, and its mouth connects to the sea. As it approaches the wilderness, it widens. As it gets nearer to the sea, it narrows into a cone.

This lake is among the wonders of the world, for twelve types of fish are found in it—one for each month—and one type of fish does not mix with the other varieties. Once a month is completed, no fish from that month's type can be found. In the following month, another variety of fish other than the first variety can be found. There is no mixing of other types of fish like this. Every month, a type of fish that has not previously been mixed is found until the year is complete. It is like this every year. These twelve types of aforementioned fish are *al-Būrī, al-Qājūj, al-Maḥal, al-Ṭalanṭ, al-Ashbalīniyat, al-Shalba, al-Qārūḍ, al-Lāj, al-Jūja, al-Kaḥlā', al-Ṭanfalū*, and *al-Qilā*.[33]

This lake connects to the south (and slightly deviating to the west) with a second lake that is called Lake Tinja. Its length is four miles and its width is the same. Between the two lakes is a mouth that connects the water between each of them. Concerning these two lakes, it is a wonder that the water of Lake Tinja is fresh and the water of Lake Bizerte is salt. One of these two lakes pours into its sister for six months. Then, their flows are reversed such that the flow of one into the other ceases, and the second lake flows into the first lake for another six months.[34] Thus, Lake Tinja does not have its waters salted by this process, nor does Lake Bizerte have its waters desalinated. This, too, is another wonder of the wonders of this land. The fish in Bizerte and Tunis are plentiful and very cheap.

From Bizerte to the city of Tabarka is seventy miles. Tabarka is a fortress on the sea with little prosperity. Around it are Arabs, who are without morals, do not retain any promises with people, and do not acknowledge their promises. Tabarka has a port for ships—and ships of al-Andalus come and use it as an anchorage. On some of the roads from Tabarka to Tunis is the city of Béja. This is a charming city built on low-lying plains that are rich in wheat and barley. It has crops that are unmatched in the

---

**32** Satfura is a region of northwestern Tunisia. Al-Idrisi's phrasing in this passage is a bit ambiguous, as he refers to first the city of Satfura and then this "splendid region." Walter E. Kaegi, *Muslim Expansion and Byzantine Collapse in North Africa* (New York: Cambridge University Press, 2010), 249–50.

**33** Bresc and Nef provide classifications of these fish in al-Idrisi, *La Première géographie de l'Occident*, 191. Those fish that are known are species of mullet, bream, porgy, sardinella, wrasse, and leerfish.

**34** The mixing of the saltwater and freshwater between these two lakes (now called the Bizerte Lagoon and Lake Ichkeul) is well known. R. H. Hughes, J. S. Hughes, and G. M. Bernacsek, *A Directory of African Wetlands* (Cambridge: IUCN, UNEP, and WCMC, 1992), 77–78; Amor Ben Garali, Mohamed Ouakad, and Moncef Gueddari, "Geochemistry and Ionic Interaction in the Bizerte Lagoon Waters (Northern Tunisia)," *Journal of Oceanography and Marine Science* 2, no. 1 (January 2011): 1–9.

Maghreb with respect to quantity and quality—this is what is found in the illustrious location of Béja. It has healthy air, much prosperity, and substantial revenues. Arabs are the collectors of the kharaj tax in Béja and its connected lands. It has a spring in its middle, to which a staircase descends, and the people of Tabarka drink from it. Outside of this spring, there is no vegetation that grows except for in fields and farms. Between Béja and Tabarka is a little more than a day's journey. Opposite Béja to the north, at the coast of the salt sea, is the city of El Kala.[35] Between the two is a long day's journey.

El Kala is a small city, which has a fortified wall and a kasbah.[36] Many Arabs are around this city. The city's prosperity is because of the fishing of coral by its people, which is available in great quantities there. It is the most splendid of all the coral that exists in other parts of the region, like the ones in the city of Ceuta or Sicily. We will mention Ceuta, which is located on a sea passage called the Strait Connected to the Sea of Darkness [*al-Zuqāq al-Mutaṣil bi-Bahr al-Ẓulamāt*].[37] Trade circulates from other lands to this city, and many people depart from it to all other destinations.

The minerals of this coral in the city of El Kala are its primary service each year. At all times, fifty boats work there (more or less), and each boat has twenty men (more or less). Coral sprouts like trees, then it becomes petrified in the same sea between two great mountains. This coral is harvested by great machines comprising many hooks that are made from hemp and are located atop boats.[38] These threads of hemp then entangle themselves around the coral so that fishermen can approach it, extract the coral, and then sell it for much money. The livelihood of the people of El Kala revolves around this. Its people drink from wells and agriculture is sparse here, though nourishment has been brought from the neighbouring bedouin Arabs. Sometimes, fruits have also been brought to El Kala from Annaba and elsewhere.

Between the city of El Kala to the city of Annaba is a short day's journey. There are twenty-four miles between their ports in a straight line. The city of Annaba is medium-sized—neither big nor small—and is located on the coast. Its extent is like that of Lorbeus. It used to have good markets, flourishing trade, and tangible profits. There used to be much good quality timber available, but now it has few orchards and trees. There are also species of fruit that are accessible for its people, and most of these fruits are from its countryside. Wheat and barley are in great quantity during the harvest, as we

---

**35** The Arabic name of this city is *Marsa al-Kharaz*, which This is literally means "Port of the Pearls." This name is likely a reference to the thriving coral industry of the city, which al-Idrisi discusses in some detail.

**36** A kasbah [*qaṣaba*] is a tower, citadel, or fortified quarter in a city. It originally connoted the idea of the "heart" of a city, but then evolved to mean a fortified central area that was typically home to the city's ruler. "Ḳaṣaba," *EI2*.

**37** This is the modern Strait of Gibraltar.

**38** These technologies are explored in Georgios Tsounis et al., "The Exploitation and Conservation of Precious Corals," *Oceanography and Marine Biology: An Annual Review* 48 (2010): 161–212; Iwasaki Nozomu, ed., *Precious Coral and the Legacy of the Coral Road* (Newcastle upon Tyne: Cambridge Scholars, 2021).

have described. In Annaba, there are also mines of good iron. These lands cultivate flax, and honey is possible to find—along with ghee. Most of the livestock in Annaba are cows. The expansive regions and earth of Annaba are located where the Arabs have supremacy. Annaba was conquered by the hand of one of the men of the great King Roger in the year 548H. Now, the city is weak and lacking in prosperity. A governor in the service of the great King Roger from the Hammad family rules over the city of Annaba.[39] Near the city is Mount Edough, which is a tall peak where there are mines of iron, which we have mentioned above.

From the city of Béja (which we have mentioned previously) to the city of Lorbeus is a two-day journey.[40] From there to the city of Kairouan is a three-day journey. Similarly, between Béja and the sea is a short two-day journey. The city of Lorbeus is located on low-lying plains and surrounded by a wall of solid earth. In the middle of the city, there are springs of running water that never become dry and from which its people drink. The name of one of the springs is the Spring of Rabah, and the other is called the Spring of Ziyad. The water of the Spring of Ziyad, which is quite healthy, is better than the water of the Spring of Rabah. Lorbeus has an iron mine, but there is not a single tree or sapling around the outside of the city. It has farms of wheat and barley, which are saved in large quantities. It is twelve miles from Lorbeus to the city of Dahmani [*Āba*], which is to the west of Lorbeus. Dahmani has saffron that compares to the saffron of al-Andalus in its quantity and quality. Indeed, the earth of these two places is one—mixed together. In the middle of the city of Dahmani is a spring of running water from which its people drink, and this spring is inundated with flowing water. In earlier times, Dahmani had a wall made of clay and prices were cheap. Now, most of the city is in ruins.

From the city of Lorbeus to the small city called Tamadit is a journey of two days.[41] Around Tamadit is an earthen wall, and its people drink from springs within it. The people of Tamadit harvest crops of wheat and barley in great quantities. Between Lorbeus and Tamadit is a small city called Marmajana.[42] Arabs have levied taxes on the people of this city. The people of Tamadit have wheat and barley that are of

---

**39** The Hammadids were the cousins and perennial rivals of the Zirid dynasty, based first in inland Qalʿa and then coastal Béjaïa (both in modern Algeria). The Normans raided or conquered a few of their coastal territories during their Ifriqiyan conquests of the 1140s. Idris, *La Berbérie orientale sous les Zīrīdes*, 1:106–19.

**40** Lorbeus is the site of a Roman settlement called Lares or Larabus, the site of which in modern Tunisia is called Henchir Lorbeus. "Lares (Henchir Lorbeus)," Trismegistos, n.d., www.trismegistos.org/place/17159.

**41** Tamadit is an Amazigh/Berber village located on the slopes of the Rif mountains. I would like to thank Nizar F. Hermes for his help in locating this settlement.

**42** Bresc and Nef associate Marmajana with "Aïn et Henchir Merdjana" near Kalaat es-Senan in modern Tunisia. Ducène argues that this site is located in the area of Thala or Bermadjenna. Although these locations are quite close to one another (about 40 kilometres), we have opted not to include them in our maps due to the uncertainty of the location of the site. Al-Idrisi, *La Première géographie de l'Occident*, 194; Ducène, *L'Afrique dans le Uns al-Muhağ Wa-Rawḍ al-Furağ d'al-Idrīsī*, 162.

sufficient quantity and excess. From Sigus [*Tīnjas*] to coastal Annaba is a three-day journey. From Sigus to the city of M'Sila is a five-day journey. Similarly, from the city of Lorbeus to Kairouan is a three-day journey. From the city of Lorbeus to Tunis is a two-day journey. From Sigus to Constantine is two days. Between Lorbeus and the city of Béjaïa is a twelve-day journey. From Marmajana to the city of Majana is a short two-day journey or a long day's journey.[43]

Majana is a small city surrounded by an earthen wall. In the past, bulbs of saffron were planted there frequently. Majana has a river of abundant water that comes from a mountain close to it, where people grow crops. This is a towering mountain, and from it are cut out millstones, which are constructed with quality and produce good flour. These millstones might even last more than the life of a human and not require engraving or maintenance because of the hardness and precision of their parts. The land of Majana was conquered by the Arabs, and they store their food in the city. Between Majana and Constantine is a three-day journey. From Constantine to Béjaïa [*Bijaya al-Nāṣiriya*] is a six-day journey.

Between Tunis and El Hamma du Jérid is a long day's journey. This journey is the width of the island that is called Bashu Island, which is a good and blessed land full of olive trees, expansive cultivated lands, blessings, bounties, and crops.[44] Water does not run much in this part of the land, but it is possible to find water in wells. The land in general is fertile enough. This island is a region that had the city of Bashu, but now nothing remains from it except for its location, an erected fort, and a fort on the sea called Nabeul. In the days of the Rum, a populous city used to be close to this fort, but it was destroyed. Now, only its site and this small fort remain. Similarly, close to the fort of Tusihan is the remains of a city that was inhabited during the days of the Rum, but it was destroyed and the only site remains.[45]

Between Tunis and the city of Kairouan is Mount Zaghouan.[46] This is a very high mountain that ships spot from far-away seas because of its height and elevation in the

---

**43** The location of Majana is also unclear. Bresc and Nef place it on the Oued Mellegue river at the confluence of the Oued Kseub river. Ducène places it at Henchir al-Hadid near Kairouan. Idris, meanwhile, thinks that it is the Tunisian village of Majen, located some 25 kilometres east of Bou Kadra in Algeria. Given the ambiguities associated with plotting this site, we have opted not to include it on our maps. Idris, *La Berbérie orientale sous les Zīrīdes*, 2:476–77; Al-Idrisi, *La Première géographie de l'Occident*, 195; Ducène, *L'Afrique dans le Uns al-Muhağ Wa-Rawḍ al-Furağ d'al-Idrīsī*, 155.

**44** It is unclear what location al-Idrisi is referring to with Bashu Island, as all of the other major islands around the coast of Ifriqiya are accounted for in the remainder of this chapter. Bresc and Nef speculate that al-Idrisi is referencing the peninsula of Cape Bon. Given al-Idrisi's association of Bashu Island with the coastal fort called Nabeul, which is located on modern Cape Bon, this speculation makes sense. Al-Idrisi, *La Première géographie de l'Occident*, 195.

**45** This fort is probably located somewhere in the modern city of Nabeul. Bresc and Nef place it on the eastern edge of Nabeul on the peninsula of Ras Maamoura. This placement makes sense given the use of the term "Cape Tusihan" as the northernmost point on the Gulf of Hammamet. See, for example, J. W. Norie, *New Sailing Directions for the Mediterranean Sea*, 2nd ed. (London: Charles Wilson, 1841), 344; Al-Idrisi, *La Première géographie de l'Occident*, 195.

**46** Mount Zaghouan is the highest point in eastern Tunisia and the source of the eponymous

air. This is the greatest of mountains with respect to water. It has fertile lands, farms, and expansive cultivation. It is inhabited by a group of unique Muslim worshippers.[47] Likewise, Mount Ousselat has a length of two days, and from it to Tunis is two days. Between it and Kairouan is fifteen miles. It has many cultivated lands and running water, as well as castles—like the castle of Almond Trees, the castle of al-Tifaf, the castle of al-Qitana, the House of Ismaʿil, and the House of Livestock [*al-Dawab*].[48] All of these lands are inhabited by the tribes of the Berbers, who are the people of this region. They live in fertile land and have livestock—cows, sheep, mules, and brood-mares. The Arabs are presently conquering the plains of this entire land.

Let us now list the roads travelled between these lands—namely that road from Kairouan to Tiaret. From Kairouan to Alaâ [*al-Juhaniyyin*], which is a village, is a day's journey.[49] To the city of Sbiba is a day's journey. This is a large, eternal city of much water and gardens. Around it is a fortified wall of stone. It also has a suburb in it with markets and inns. The people of Sbiba drink from a large, running spring, around which are gardens, orchards, and crops of cumin, caraway, and legumes.

From Sbiba to Marmajana, which is a village of the Hawwara, is a day's journey.[50] From there to the city of Majana, which we have previously mentioned, is a day's journey. Then, to Meskiana is a day's journey. This is an old and populous village with crops, profits, and springs. It has a market—bigger than Marmajana—that extends in a line like a string. From Meskiana, one comes to Baghai, which is a populous city that we have previously mentioned in this chapter. The road carries on from the city Baghai to M'Sila, as we previously mentioned.

A second road goes from Kairouan to M'Sila that, in contrast to the road we have previously mentioned, exits from Kairouan to Aïn Djeloula in a short day's journey. Aïn Djeloula is a small city with a wall around it. It has a spring of running water with many orchards and date palms. From there to Ahara is a day's journey.[51] This is a charming village, the water of which comes from wells. It also has many crops of wheat and barley.

From Ahara to the village of Tamajina is a day's journey.[52] It has a large tract of land on which is grown wheat and barley, which may be sold very cheaply. From there to the city of Lorbeus is a day's journey. From Lorbeus to Tifash is a day's journey.[53]

---

aqueduct that brought water to the city of Carthage—a distance of over 100 kilometres.

**47** It is unclear who these worshippers are—perhaps an offshoot of the Ibadis or Kharijites whose practices deviated from these (already marginal) groups?

**48** The locations of these five sites are unknown.

**49** The association of al-Juhaniyyin with modern Alaâ is provided by Bresc and Nef. Al-Idrisi, *La Première géographie de l'Occident*, 196.

**50** The Hawwara is a Berber tribal confederation that had a substantial presence in Ifriqiya and Egypt during the Middle Ages. "Hawwāra," *EI2*.

**51** Ahara is the site of ancient Aggar, the ruins of which are now called Henchir Sidi Amara. Al-Idrisi, *La Première géographie de l'Occident*, 197.

**52** The location of Tamajina is unknown.

**53** Tifash is located on the ancient site of Tipasa in inland Algeria (not to be confused with modern

This is also an ancient, eternal city around which is an ancient wall made of stone and lime. It has a spring of running water as well as orchards, meadows, and crops of mostly barley.

From Tifash to the fort of al-Ifriqi is a day's journey.[54] It does not have a wall, but it has farms and substantial investments in wheat and barley. From there to the village of Ain Arko [*Azkū*] is a day's journey. It has gardens, springs, water, orchards, crops of wheat and barley, and wide-ranging wealth. From there to the village of al-Bardawan is a day's journey. It used to be a large village. It is among the regions of wheat and barley.[55]

From al-Bardawan to the village of al-Nahrin is a day's journey. It is located on low-lying plains, and in it are wells of freshwater. It used to have a market. The majority of the people who live there are Berbers from the Kutama and Mazata tribes. From there to the village of Tamasit of the Kutama, which has trees and cultivated fields, is a day's journey.[56] From there to Dakama is a day's journey.[57] It is a village with a market and people of the Kutama tribe. From there to El Hammadia [*Awsaḥanat*] is a day's journey. This is a village of the Berbers with running water and farms of wheat and barley. From there to M'Sila is a journey of less than a day.

From the city of M'Sila to Ouargla is a long twelve-day journey. This city is inhabited by rich tribes and wealthy merchants who roam across the lands of the Blacks to the lands of Ghana and the lands of Wangara to extract gold ore from there.[58] Then, they mint the gold in their lands in the name of their land. These people are Wahbiyya Ibadis and abominable Kharijites outside of the religion of Islam.[59] From Ouargla to

---

Tipaza on the coast of modern Algeria). Cambuzat, *L'évolution des cites*, 2:223; al-Idrisi, *La Première géographie de l'Occident*, 197.

**54** Per the work of Cambuzat, the fort (*qaṣr*) of al-Ifriqi is the modern site of Ksar al-Frigui, which is located northeast of modern Sedrata. Cambuzat, *L'évolution des cites*, 2:218. This site is identified as location 376 in map 18 (Souk-Arrhas) of Gsell, *Atlas archéologique de l'Algérie*.

**55** The location of al-Bardawan is unknown.

**56** The locations of al-Nahrin and Tamasit are unknown.

**57** Bresc, Nef, and Cambuzat place Dakama near the modern site of Aïn Oulmene in inland northeastern Algeria. Cambuzat, *L'évolution des cites du Tell*, 2:79–81; al-Idrisi, *La Première géographie de l'Occident*, 197.

**58** The exchange of gold, spices, slaves, and salt is documented in the eleventh and twelfth centuries, though much of the intercultural encounters that happened along these trans-Saharan trade networks are now lost. The most revealing textual source from the Middle Ages about trans-Saharan trade is the travelogue of Ibn Battuta, which is translated in Levtzion and Hopkins, *Corpus of Early Arabic Sources for West African History*, 279–304. See also Michael Brett, "Ifriqiya as a Market for Saharan Trade from the Tenth to the Twelfth Century AD," *The Journal of African History* 10 (1969): 347–64; E. Ann McDougall, "The Sahara Reconsidered: Pastoralism, Politics and Salt from the Ninth through the Twelfth Centuries," *African Economic History* 12 (1983): 263–86; Sarah M. Guérin, "Forgotten Routes? Italy, Ifrīqiya and the Trans-Saharan Ivory Trade," *Al-Masāq: Islam and the Medieval Mediterranean* 25 (2013): 70–91; François-Xavier Fauvelle, *The Golden Rhinoceros: Histories of the African Middle Ages*, trans. Troy Tice (Princeton: Princeton University Press, 2018).

**59** The Ibadis are a Muslim sect that had a substantial presence in medieval Ifriqiya, particularly in inland areas and the island of Djerba. The Wahbiyyas and Kharijites (also called the Nukkaris)

Ghana is a thirty-day journey. From Ouargla to Kugha is around a month and a half. From Ouargla to Gafsa is a thirteen-day journey.[60]

Now, let us return to an account of the city of Gabès, which is on the coast of the salt sea. Gabès is a city of the Africans, who were previously mentioned. From the city of Gabès to al-Fawara, which in earlier times was a village but is now in ruins, is thirty miles. From there to the wells of Khabat is thirty miles.[61] From there to the fort of al-Daraq is twenty-eight miles.[62] From there to the well of the Camel Drivers [*al-Jamālīn*] is thirty miles.[63] From there to Sabratha is twenty-four miles.[64] From the fort of Sabratha to Tripoli is a day's journey. All of these residences that we have mentioned on this road are uninhabited and deserted. The Arabs attacked their cultivated lands and obliterated their vestiges—their wealth ruined and their bounties erased. Now, these lands have no amiable inhabitants nor allied residents. They have been taken over by Arab tribes called Mirdas and Riyah.

Another road is from Gabès to the river of Ahnas, then to the well of the Zanata, then to Tamdafit, to the wells of al-ʿAbbas, to Tafnat, to the well of al-Safȧ, and to Tripoli.[65] The city of Tripoli is a fortified city, surrounded by a wall of stone. Located on the coast, it is white, has beautiful streets, and contains tidy markets. It has factories and warehouses from which people bring goods to various places. Previously, there was a surplus of production from all the lands near the city: many olive and fig trees, plentiful fruits, and dates. However, the Arabs damaged the city and the places around it. They evacuated its people, its countryside was abandoned, and its situation was changed. Trees were annihilated and waters ran dry. The great King Roger conquered Tripoli in the year 540H, taking its women prisoner and killing its men.[66] Now, the

---

are sects within this movement. The latter group was notorious (in the eyes of later authors) for their revolt against Fatimid rule in the tenth century and for their later pirate activities at Djerba. Writers like al-Idrisi perceived the Ibadis to be heretical. Valerie Hoffman, *The Essentials of Ibadi Islam* (Syracuse: Syracuse University Press, 2012); Cyrille Aillet, "Wārjlān, un foyer de l'ibadisme médiéval aux marges du Sahara," in *L'ibadisme dans les sociétés de l'Islam médiéval*, ed. Cyrille Aillet (Berlin: De Gruyter, 2018), 207–43; Paul M. Love Jr., *Ibadi Muslims of North Africa: Manuscripts, Mobilization, and the Making of a Written Tradition* (Cambridge: Cambridge University Press, 2018); Tarek Kahlaoui, "Ibadi Jerba: Surviving the Early Ḥafṣids (13–14th Centuries)," *The Journal of North African Studies* 26 (2021): 631–41.

**60** These regions are considered in greater detail in the relevant sections of the first three sections of the First (southernmost) Clime of al-Idrisi. The modern locations of Ghana (the city) and Kugha are unknown.

**61** The locations of al-Fawara and the wells of Khabat are unknown.

**62** Bresc and Nef place the fort of al-Daraq at the site of Gsir el-Dreg, which is near modern Zelten. Al-Idrisi, *La Première géographie de l'Occident*, 198. See also Idris, La *Berbérie orientale sous les Zīrīdes*, 2:460; Henri Bresc, "La Sicile et l'espace Libyen au moyen-âge," *Africa: Rivista trimestrale di studi e documentazione dell'Istituto italiano per l'Africa e l'Oriente* 63 (2008): 187–208.

**63** The location of the well of the Camel Drivers is unknown.

**64** Al-Idrisi lists both Mansouria and Sabratha as *Ṣabra* in his work.

**65** The locations of all of these sites between Gabès and Tripoli are unknown.

**66** On Tripoli and the Norman conquest of the city, see Michael Brett, "The City-State in Medieval

city remains loyal to him and is counted among all his lands. The land of the city of Tripoli is not comparable with respect to the cultivation in agriculture. There is nothing like it known in any other lands, which is well known.

From the city of Tripoli eastward to the city of Sirt is two hundred and thirty miles, which is an eleven-day journey.[67] Also, when a traveller departs from the city of Tripoli to al-Majtana, it is twenty miles.[68] From there to Wardasa is twenty-two miles. From there to Raghugha is twenty-five miles.[69] From there to Tawergha is twenty-two miles. Then, to al-Mansaf is twenty-five miles.[70] Then, to the forts of al-Hassan ibn al-Nuʿman al-Ghassani is forty miles.[71] Then, to al-Asnam is thirty miles.[72] Then, to Sirt is forty-six miles. This road is far from the coast sometimes and close to it at others—and all of this territory is under the control of two tribes from the Arabs, the ʿAwf and the Dubbab.

Between the city of Sirt and the sea is two miles. Around Sirt is an earthen wall, which is further encircled by sand. In the city are remains of date palms (but no olive trees) as well as many mulberry trees and fig trees—though less since the Arabs arrived and devastated much of them. Sirt does not have any grass like Awjila, nor does it have dates like Waddan.[73] In the past, date palms used to be more than sufficient for subsistence. The people of Sirt used to have grapes and fruits, though they have spoiled in our time. Nothing remains of them except what was in the depths of the valleys and the tops of the mountains. The water from rain comes into cisterns and its wells are few. Around Sirt are the tribes of the Berbers.

Near the city of Tripoli is Mount Maqada, and between the two is a three-day journey. From the city of Tripoli to Mount Nafusa is a six-day journey. From there to Sfax is a nine-day journey. From Mount Nafusa to Qastiliya is a six-day journey. The people of Mount Nafusa are all Muslim, but are Kharijites who deny the doctrine of Ibn

---

Ifriqiya: The Case of Tripoli," *Cahiers de Tunisie* 34 (1986): 69–94.

**67** The mediaeval city of Sirt was located some 55 kilometres east of the modern city at a site that is now called al-Mudayna.

**68** The location of al-Majtana is unknown.

**69** The locations of Wardasa and Raghuga are unknown, though there is a Wadi Wardasa in Libya east of modern Sirte that could roughly correspond to this location. Bresc and Nef speculate that Raghuga could be located at the oasis of Zliten in Libya. Al-Idrisi, *La Première géographie de l'Occident*, 199.

**70** The location of al-Mansaf is unknown.

**71** We follow Bresc and Nef by placing this site at Thamad Hasan in Libya. Al-Idrisi, *La Première géographie de l'Occident*, 199; *Libya, Official Standard Names Approved by the United States Board on Geographic Names*, 2nd ed. (Washington, DC: Defense Mapping Agency Topographic Center, 1973), 286.

**72** Bresc and Nef place al-Asnam at Marsa Zafran, which corresponds to modern Sirte. Al-Idrisi, *La Première géographie de l'Occident*, 199; Jacqueline Passon et al., "Places of Trade, Communication and Religion—Important Oases," in *Across the Sahara: Tracks, Trade and Cross-Cultural Exchange in Libya*, ed. Klaus Braun and Jacqueline Passon (New York: Springer Open, 2020), 165–247.

**73** Awjila and Libya are oasis towns in southern Libya.

Munabbih al-Yamani.[74] We have already mentioned this doctrine in our discussion of the people of the island of Djerba.[75]

From Mount Nafusa to Mount Demmer is a three-day journey across unceasing sand. On the peripheries of this mountain is a group from the Berbers called Rahana. They are a group that breeds camels and rides upon the strongest and speediest of them. Then, the Rahana march in groups to Arab tribes (settled far from them) to strike, attack, and take their camels before returning with their spoils/booty to the mountain and their dwellings where they take refuge. They do not have any occupation besides this. Not one of the Arabs neighbouring them does not complain about their harm. It is rare to catch one of them due to the running speed of their bred camels, their understanding of that land, and the fortifications of their land (as we have said). These lands are connected to the south with the lands of Waddan.

Now, we recall what comprises this chapter—which we are describing—with regard to ports of the sea, capes, inhabited forts, and lands calculated according to what was asked, researched, reached by force and energy, and the guidance of Allah.

Thus, I say that from the western city of Annaba to its cape is six miles. Then it is forty miles to the bay of al-Azqaq. This is a small bay that ends at El Kala, which has a cape that extends into the sea.[76] From El Kala to Tabarka is twenty-four miles. From there to the edge of the bay is fifteen miles straight across and twenty-four miles along its coast. There is a beach called the Saw [*al-Minshār*] that is sixteen miles long. From its edge to the stronghold of Abu Khalifa is ten miles.[77] From there, cutting across the bay by sea is twenty miles, while going along the coast is twenty-eight miles. It is then twelve miles to the edge of the cape. From there to Bizerte, which we have already mentioned, is eight miles. From Bizerte to the port of the Banu Wajas is twelve miles.[78]

From the cape of the Banu Wajas to the edge of the mountain is thirteen miles via the bay, along which are various forts. From the beginning of the edge of Banu Wajas to the fort of the Port of the River [*al-Wadī*], which is the location of a small river, is three miles.[79] From there to the fort of Tarsha Dawud is three miles.[80] From there to the fort of Sounine is five miles. From there to the edge of the mountain, which is also

**74** Wahb ibn Munabbih was a prolific Muslim author of the seventh and eighth centuries. "Wahb b. Munabbih," *EI2*.

**75** This passage is curious, for al-Idrisi has not yet described the people of Djerba. Perhaps during the course of the compilation of this geography, al-Idrisi modified the order of presentation of these Ifriqiyan sites.

**76** This is the modern Bay of Annaba, which extends in the west from Cape Garde to the eastern Cape Rosa.

**77** The location of the stronghold of Abu Khalifa is unknown.

**78** The location of this port is unknown, though presumably it was on the coastline directly east from Bizerte on modern Cape Zebib.

**79** The location of this fort is unknown, but it is likely in the vicinity of modern Metline in Tunisia.

**80** The location of the fort of Tarsha Dawud is unknown, but (given the known location of Sounine), it makes sense as being somewhere in the area of modern Ras Jabal.

known as the Church [*al-Kanīsa*] is two miles. This is the head of the bay in the middle of which is the city of Tunis and its lake.

From the edge of this mountain (via its curved base) to the site of the river of Medjerda is six miles. From the site of the river to the fort of Jilla, which is close to it, is around four miles.[81] From there to the fort of Jardan is two miles.[82] From there to the city of Carthage is two miles. The city of Carthage is in ruins, as we have previously mentioned.

From Carthage to the throat of the river of Tunis, which cuts its bay in half, is three miles. From the throat of the river to the fort of Jaham is twelve miles.[83] From the fort of Jaham to the fort of Korbous is sixteen miles. From the fort of Korbous to the edge of Ifran, which is a cape that extends into the sea, is fourteen miles. The coast of this bay is seventy-four miles, but cutting across it in a straight line from the edge of the mountain to the edge of Ifran is twenty-eight miles. Likewise, from the middle of the bay (where the mouth of the river of Tunis is located) to the edge of Ifran is twenty-eight miles in a straight line, but going along its curve is fifty-six miles.[84]

From the edge of Ifran to the port of the fort of the Date Palm [*al-Nakhla*] is six miles. From there to the fort of Minzit is twelve miles. From there to the fort of Nuba is thirty miles.[85] In all, from the mouth of the river of Tunis all the way to Nuba is seventy miles. In front of Nuba, in the sea, there are the islands of Zembra [*al-Jāmūr al-Kabīr*] and Zembretta [*al-Jāmūr al-Saghīr*]. Between the two islands is seventy miles. From Zembra to Nuba is twelve miles. From Nuba to the edge of al-Rakhima is one mile cutting across the bay, and this bay along its edges is six miles—and all of it is short.

From the edge of al-Rakhima, one finds Cape al-Baqla, which is the edge of the mountain called Adaru, which is on the side of Kelibia to the east.[86] From Cape al-Rakhima to Zembretta is six miles. Zembra and Zembretta are two mountains rising from the sea that are used as anchorages from violent winds. It is thirty miles in total between Nuba and Kelibia. From the edge of Kelibia to Monastir is a waterway.

---

**81** Bresc and Nef place Jila at the site of Gella in Tunisia. Al-Idrisi, *La Première géographie de l'Occident*, 201.

**82** Bresc and Nef place the fort of Jardan at the site of ancient Utica, which (today) is located some 11 kilometres inland. The shoreline of this area, though, has changed substantially since the Middle Ages due to modern land reclamation projects. During the ancient and mediaeval periods, the area around Utica was located at the mouth of the Medjerda River. Al-Idrisi, *La Première géographie de l'Occident*, 201; Hugo Delile et al., "The Geoarchaeology of Utica, Tunisia: The Paleogeography of the Mejerda Delta and Hypotheses Concerning the Location of the Ancient Harbor," *Geoarchaeology* 30 (2015): 291–306.

**83** The fort of Jaham is likely the site of Sidi Jammi in modern Hammam-Lif, which is on the coast southeast of Tunis. Al-Idrisi, *La Première géographie de l'Occident*, 202.

**84** In modern parlance, this area is called the Gulf of Tunis, which extends from Sidi Bou Said in the west to Ras al-Fortas in the east.

**85** The location of these three sites is unknown, but presumably they are located on the northern coast of Cape Bon. Given the proximity of Nuba to the islands of Zembra and Zembretta, perhaps it makes the most sense to place it on Ras al-Ahmar.

**86** Cape al-Baqla is likely the tip of Cape Bon.

Someone who walked from Kelibia to the fort of Abu Marzuq would travel seven miles.[87] From there to the fort of Lebna is eight miles. From Lebna to the fort of Saʿd is four miles. From the fort of Saʿd to the fort of Korba is eight miles and to the edge of Tusihan is ten miles. The edge of Tusihan extends into the sea for a mile and a half—like a protruding molar. From this molar to the fort of Tusihan on the bay is four miles.[88]

From Tusihan to the fort of Nabeul is eight miles. Nabeul was a very large and populous city of the Rum. When the island was conquered at the dawn of Islam, its virtues and charms were plundered until no vestige or remnant of it remained except for only the site of the fort. The remains of the ruins of Nabeul indicate its location. From the fort of Nabeul to the fort of the Tailor [*al-Khayyāṭ*] is eight miles. Between it and the sea is around two miles. From the fort of the Tailor to the fort of the Date Palm [*al-Nakhīl*] is six miles.[89] Then, to the edge of El Hammamet is seven miles. From this edge back across land to the city of Tunis is a long day's journey, which is the width of the island called Bashu Island, which was mentioned previously. This cape is called Cape Hammamet. It has a fort built on the cape, which extends into the sea for about a mile.

From Hammamet to the Lighthouse [*al-Manār*], which is a fort that is far from the sea, is five miles.[90] From there to the fort of the Sentry [*al-Marṣad*] and then to the fort of the Almoravids is six miles.[91] The fort of the Almoravids is at the bottom of the Bay of the Buried [*al-Madfūn*]. From there to the edge of the Cape of the Buried [*al-Madfūn*] is six miles.[92] From the edge of the aforementioned cape to the castle of Hergla is eight miles.

From Hergla to the city of Sousse is eighteen miles. This is a city full of many people and shops. Travellers strive to go there and export their commodities, which are not comparable with regard to types of attire and suitable turbans. These are good and valuable commodities. Sousse has bustling markets, its water is from cisterns, and around it is a wall of fortified stone. From Sousse to the fort of Skanes is eighty miles. From Skanes to the fort of Ibn al-Jaʿd is four miles.[93] From there to the forts of Monastir is two miles. In all, from the castle of Kelibia to Monastir when cutting across

**87** The location of the fort of Abu Marzuq is unknown, but is likely in the vicinity of modern Manzil Tamim, which is located about halfway between Kelibia and Lebna.

**88** The location of this site is unknown but is in the area of modern El Maâmoura.

**89** The locations of the fort of the Tailor and the fort of the Date Palm are unclear.

**90** The modern rendering of these ruins is Ksar Menara.

**91** The locations of these sites are unknown. The Almoravids were a dynasty based in modern Morocco that were ousted by the Almohads in the twelfth century. Some factions within the Almoravids migrated east to Ifriqiya; presumably this fort is one such fortification set up by displaced Almoravids. Amar Baadj, *Saladin, the Almohads and the Banū Ghāniya: The Contest for North Africa (12th and 13th Centuries)* (Leiden: Brill, 2015).

**92** The modern Forêt el-Madfoun, which is directly northwest of Hergla, retains this mediaeval name.

**93** The location of this fort is unknown.

in a straight line is one hundred miles, but when travelling along curved coastal waters it is one hundred and twenty miles.

Al-Monastir is directly opposite the sea from the island of Kuriat. From there to Monastir is nine miles. From this island to Lamta is ten miles. From there to the Dungeon [*al-Dīmās*] is twelve miles.[94] From there to Mahdia is twenty miles. Similarly, from Monastir to Mahdia is also thirty miles. From Monastir to the fort of Lamta is seven miles. From the fort of Lamta to the Dungeon is eight miles. From the Dungeon to Mahdia is eight miles.[95]

Mahdia is surrounded by the sea, as we have previously mentioned. From there, the sea begins to curve toward the south. From Mahdia to the fort of Salakta is six miles. From there to the fort of Elevation [*al-ʿĀliya*] is six miles, and then to Kaboudia is thirteen miles.[96] Kaboudia is a charming fort where fish are caught (all of them extraordinary) and sold for very cheap. From Kaboudia to the fort of Melloulèche [*Maliyan*] is eight miles. From the fort of Melloulèche to the fort of Basil [*al-Rayḥāna*] is four miles, and then to the fort of Qanata is four miles.[97] At the fort of Qanata is worked much unsophisticated pottery, which is prepared for Mahdia and elsewhere. Its clay is red.

Then, to the fort of the Almond [*al-Lawza*] is four miles and next, to the fort of Ziyad is six miles.[98] From the fort of Ziyad to the fort of Majdunis is eight miles. From the fort of Majdunis to the fort of Qasas is eight miles. From the fort of Qasas to the fort of Qazil is two miles. In all, from the fort of Ziyad to Cape Qazil is eighteen miles. From Cape Qazil to the fort of Habala is two miles on the bay.[99] From there to the city of Sfax is five miles on the bay. Considering all of this, from Kaboudia to Sfax is forty-eight miles along the coast and thirty miles cutting across it in a straight line.

Near the fort of Ziyad in the sea toward the east is the island of Kerkennah—a place that is located between the fort of Ziyad and Sfax.[100] It is twenty miles from Kerkennah to Kaboudia. From Kerkennah to Sfax is around fifteen miles. Kerkennah is a charming island and well-populated. There is no city there, but the people live peacefully in huts. The island is a fertile land and has many vineyards, grapes, harvests of cumin and

**94** The modern location of this site is likely Rass Dimasse (sometimes rendered as "Dimass").

**95** The crisscrossing routes that al-Idrisi provides for the sites of Monastir, Lamta, the Dungeon, and Mahdia indicates its importance to the Normans, who conquered Mahdia in 1148 and suffered a defeat to the Zirids and their allies at the Dungeon in 1123.

**96** The fort of Elevation corresponds to the marina of El Alia between Salakta and Kaboudia.

**97** Bresc and Nef speculate that the fort of Basil is likely the site of Henchir Karaa south of Melloulèche and the fort of Qanata is the site of nearby Henchir bou Tria. Al-Idrisi, *La Première géographie de l'Occident*, 204.

**98** The fort of the Almond corresponds to modern La Louza. The locations of the fort of Ziyad and all of the other locations through the fort of Qazil are unknown.

**99** Bresc and Nef place the fort of Habala at the site of Henchir Jebla near modern Sfax. Al-Idrisi, *La Première géographie de l'Occident*, 204.

**100** The Kerkennah Islands are a chain of islands located off the coast of Sfax in modern Tunisia, though al-Idrisi treats them as a singular entity.

anise (the sweet seed). The great King Roger conquered this island in the year 548H. On the western side of the island, caves and caverns provide security for those who escape to them. They are called al-Qarbadi and they connect to the Short Stone [*Ḥajr Qaṣīr*] for twenty miles. From al-Qarbadi to the House of the Short [*Bayt al-Qaṣīr*] is 35 miles.[101] The length of this island is sixteen miles and its width is six miles.

Next, we return now to the aforementioned Sfax to say that from it to the Cape of Sand [*al-Ramla*] is four miles. From the Cape of Sand back to the south (where the bay begins) to the fort of the Magi is four miles.[102] From there to the fort of Banaqa is ten miles.[103] From the fort of Banaqa to the fort of Tanidha is eight miles.[104] From there to the forts of the Rum is four miles.[105] From there to the city of Gabès, which we have already described and mentioned what is its nature, is seventy-five miles.

Then, from Gabès along the coast to the fort of Ibn ʿAyshun is eight miles.[106] To the fort of Zajuna, it is eight miles.[107] From the fort of Zajuna to the fort of the Banu Måmun is twenty miles.[108] From the fort of the Banu Måmun to Amrud is eleven miles.[109] From there to the fort of the Cliff [*al-Jurf*] is eighteen miles.[110] In all, from the tip of the Cape of Sand to this edge, which is called the Cliff of Abandonment [*al-Takhliya*] is fifty miles across the desert. Along the coast, it is one hundred and fifty miles.

From the Cape of the Cliff to the island of Djerba by sea is four miles. This island is populated by tribes from the Berbers.[111] Their brown skin has triumphed over the

**101** It is unclear what kinds of sites al-Idrisi is evoking in this description—presumably these were settlements distinct from the caves and caverns where people retreated during the Norman conquest?

**102** The location of the fort of the Magi is unknown.

**103** Bresc and Nef place the fort of Banaqa at the ruins of ancient Onga/Younga. Al-Idrisi, *La Première géographie de l'Occident*, 205.

**104** Per Bresc and Nef, the fort of Tanidha is located on the site of Bir Touira near modern Skhira. Al-Idrisi, *La Première géographie de l'Occident*, 205.

**105** The location of the forts of the Rum (likely meaning forts built by the Byzantines or Romans) is unknown.

**106** The site of the fort of Ibn ʿAyshun is located at the ruins of Henchir Aichoun (in modern parlance), located near the Oued el Hachchane southeast of Gabès.

**107** Bresc and Nef place the fort of Zajuna at the ruins of Henchir Dassour near Zarat Beach in Tunisia. Al-Idrisi, *La Première géographie de l'Occident*, 205.

**108** The location of the fort of the Banu Måmun is unknown.

**109** Bresc and nef place Amrud at Garaet el Hamroun. Al-Idrisi, *La Première géographie de l'Occident*, 205.

**110** The precise location of the fort of the Cliff is unknown, but it likely corresponds to somewhere in the area of Al Jorf in modern Tunisia, which borders the island of Djerba.

**111** The presence of Ibadi Muslims, who were seen as heretical by Sunni Muslims, is well attested on the island of Djerba. See, for example, Virginie Prevost, "La chaussée d'al-Qantara, pont entre Djerba et le continent," in *Autour de la géographie orientale… et au-delà. En l'honneur de J. Thiry.*, ed. L. Denooz and X. Luffin (Louvain: Peeters, 2006), 165–88; Paul M. Love Jr., "An Ibadi Islandscape: Ibadi Communities on Djerba in the Medieval Period," in *L'ibadisme Dans Les Sociétés de l'Islam Médiéval*, ed. Cyrille Aillet (Berlin: De Gruyter, 2018), 190–206.

other colours of the island's people. Evil and hypocrisy exist in their characters, and their Berber language is distinct and public. The people are seditious and prone to renouncing their allegiances. The great King Roger conquered this island with a fleet that he sent to the island at the end of the year 529H. Then, the people settled down and continued their lives until the year 548H, when they rose up and renounced their allegiance to the great King Roger, who launched a fleet against them in this year. He conquered the island a second time, reducing all to slavery in Palermo. The length of the island of Djerba is sixty miles from west to east. The width of the eastern side is fifteen miles. From this cape to the mainland is twenty miles. This narrow side of the island is called Cape Karin and the wide side is called Antijan.

The island of Zarzis [*Zīzuwa*] connects to Djerba to the east.[112] Zarzis is very small, but has date palms and vineyards. Between the island of Zarzis and the mainland is around a mile. The fort of the Banu Khatab is opposite it. This island is full of people who are part of the abominable Kharijites—of the Wahbiyya doctrine. As such, all of the people in the castles and forts that exist on these two islands follow it. They do not clean the garment of a stranger, nor do they touch it with their hand. They do not eat with a stranger; they do not even let them eat from their vessels unless they have been properly stored. No one gets close to their vessels except for their men and women, who purify themselves with them every day in the morning. They perform ablutions, then the Tayammum for each prayer.[113] If a passer-by draws any water from their wells, and they see it, they will banish him and then extract that water from the well. The garments of the impure are not approached by the pure, nor are the clothes of the pure approached by the impure. Despite this, these people are all hospitable by providing food and distributing it. They make peace with regard to their money, and they provide evidence-based justice for those who reside among them. From the edge of the island (I mean Djerba) that is called Antijan to Short al-Bayth is ninety miles.[114] Also, from the edge of Antijan to the bridge that is at Kerkennah is sixty-two miles.

Returning to what we were saying—to the aforementioned Cape of the Cliff. From there to the Cape of the Rivers [*al-Awdiya*] along the coast is twenty-four miles. From there to the forts of al-Zarat is twenty miles.[115] These three forts are located on the coastline facing the island of Djerba, and between them by sea is twenty miles. From the forts of al-Zarat to the fort of the Banu Dhakumin is twenty-five miles. From the Banu Dhakumin to the fort of the Granary [*al-Hara*] is six miles.[116] From there to the

**112** We follow the logic of Bresc and Nef, who see this as not an island, but as the region around modern Zarzis. Al-Idrisi, *La Première géographie de l'Occident*, 206. On this island, see Virginie Prevost, "Zîzû, l'île Mystérieuse d'al-Idrîsî," *Acta Orientalia Belgica* 18 (2005): 323–38.

**113** The Tayammum is an Islamic ritual purification done before prayer. "Tayammum," *EI2*.

**114** It is unclear what kind of settlement (if any) Short al-Bayth (*Qaṣīr al-Bayth*) is meant to indicate. Bresc and Nef take this term as a scribal error for what should have been *Qaṣīr al-Bayt*, "the fort of the house." Al-Idrisi, *La Première géographie de l'Occident*, 206.

**115** Bresc and Nef indicate that the forts of al-Zarat correspond to the ancient site of Zitra. Al-Idrisi, *La Première géographie de l'Occident*, 207.

**116** The locations of the fort of the Banu Dakhumin and the fort of the Granary are unknown.

fort of Jarjis is six miles.[117] From the fort of Jarjis to the fort of the Banu Khatab is twenty-five miles. The fort of the Banu Khatab is on the end of the Swamp of the Dogs [*al-Kilāb*] to the west, and it is opposite the island of Zarzis by sea.[118] Zarzis has a length of forty miles and its width is around half a mile. Some of it is populated with palm trees and orchards, while the remainder of it is under the water (as we previously mentioned), which is lower than the surface by a fathom (more or less). From the fort of the Banu Khatab to the fort of Shamakh is twenty-five miles.[119] Between them is a small bay called the Bay of the Donkey's Backbone [*Ṣulb al-Ḥimār*].

From the fort of Shamakh to the fort of Salih is ten miles.[120] The fort of Salih is located on a cape that extends from the east to the west. Its length is five miles and it is called the Cape of the Bakery [*al-Makhbaz*]. From there to the fort of Kutin is twenty miles. From the fort of Kutin to the fort of the Banu Walul is twenty miles. From the fort of the Banu Walul to the port of Markiya is twenty miles. From the fort of Markiya to the fort of ʿAfsalat is twenty miles. From the fort of ʿAfsalat to the fort of Sarba is four miles. From there to the fort of Sinan is two miles. From there to the fort of al-Bandari is three miles. Then, to the fort of Gharghara is ten miles. From the fort of Gharghara to the fort of the Hunter [*Sayad*] is six miles.[121] Then, to the city of Tripoli is twenty miles. We previously described the city of Tripoli and what has transpired in this research—as well its description and situation.

From the city of Tripoli to the fort at Cape Qaliyusha is fourteen miles. From there to the fort of the Book [*al-Kitāb*] is eight miles. From there to the fort of the Banu Ghasan is twelve miles.[122] To the mouth of the river of Ladis is eighteen miles. From there to the edge of the Cape of the Woods [*Shaʿrā'*] is fourteen miles. In all, from Cape Qaliyusha to the Cape of the Woods by sea is forty miles and along the coast is fifty-two

**117** Although the location of these forts is unclear, they likely correspond broadly to the modern city of Zarzis, which presumably derives from the Arabic transliteration for Gregory (the Patrician). See note 28 in this chapter.

**118** Al-Idrisi is referencing here the Bahiret el Bibane, which is a coastal lagoon in northeastern Tunisia. Although the site of the fort of the Banu Khatab itself is unknown, it was likely in the vicinity of modern Jedaria on the western edge of this lagoon. Perhaps the location of modern Ra's Ajdir would make more sense here?

**119** The location of the fort of Shamakh is unclear, for even though there is a modern city called Sidi Chemmakh in modern Tunisia, its location (both inland and northwest of the most recent sites considered in al-Idrisi) make it an unlikely candidate for al-Idrisi's "Shamakh." Instead, the area around modern Ra's Ajdir is preferable.

**120** The location of the fort of Salih is unknown, though it would make sense to be located on Farwa Island, which extends east–west slightly off the coast of modern Libya.

**121** Bresc and Nef provide the following locations for the sites mentioned in this paragraph: the port of Markiya is the abandoned site of Marsa Markan; the fort of ʿAfsalat is the site of the ruins of Gasr Tellil; the fort of Sinan corresponds to the modern city of Surman; the fort of Gharghara is located at modern Gharghuza; the fort of the Hunter is placed at modern Sayyad. The locations of the fort of Kutin, the fort of the Banu Walul, the fort of Sarba, and the fort of al-Bandari. Al-Idrisi, *La Première géographie de l'Occident*, 207.

**122** The location of these two forts is unknown.

miles. From the Cape of the Poets to the fort of Sharikas is four miles.[123] To Cape al-Masin, which is a cape extending into the sea, is four miles.[124] From there to Lebda is four miles.

The city of Lebda used to have many cultivated fields that included all bounties.[125] It is located far from the sea, so the Arabs oppressively controlled it and its lands. Thus, those things that Lebda had among its comforts were changed and its people fled for other places, so its remains are only two large forts and their accompanying cultivated fields. The residents of these forts are a group from the Hawwara Berbers. Along the coast, there is also a large fort full of people, industries, and a bustling market. At Lebda, there are many date palms and olives, the oil of which is extracted in season.

From Lebda to the fort of the Banu Hasan is seventeen miles.[126] From there to the port of Bakruwa, which is a good port that does not have any winds, is one mile.[127] From there to the fort of Hashim and to the fort of Samiya is twelve miles.[128] From the fort of Samiya to the little market [*suwayqa*] of Ibn Mathkud is twelve miles.[129] From Suwayqa to the famous Cape Qanan is twenty miles. In all, from Tripoli to Cape Qanan is one hundred and eighty miles across the desert and along the coast is two hundred and ten miles.

Here, in this chapter, the researched account of the coast of the Mediterranean has finished as ordered by the division of this book. We will come to an account of what remains from it in what comes next by the will of Almighty Allah. The little market that is attributed to Ibn Mathkud that we previously mentioned is inhabited, in and around it, by tribes from the Hawwara Berbers under the authority of the Arabs. It has a well-known market and many forts. Its people cultivate barley using irrigation, and the Arabs hoard their food.

Thus, we end the Second Part of the Third Clime—praise be to Allah! The third part follows it, Allah willing.

---

**123** The location of the fort of Sharikhas is unknown.

**124** This is modern Rás el-Msenn, located northwest of modern Khoms in Libya.

**125** Lebda is the modern site of Leptis Magna, which was a prominent city in the Carthaginian and Roman empires.

**126** The location of the fort of the Banu Hasan is unknown.

**127** The likely site of the port of Bakruwa is modern Marsa Ugra, which is located north of modern Khoms. *Libya, Official Standard Names Approved by the United States Board on Geographic Names*, 673; Al-Idrisi, *La Première géographie de l'Occident*, 208.

**128** The location of the fort of Hashim is unknown, but the fort of Samiya likely corresponds to the ruins of Bir Shumiya, located in the oasis of Tawergha. Al-Idrisi, *La Première géographie de l'Occident*, 208.

**129** Per Bresc and Nef, a suwayqa is a "little market." The location of this market of Ibn Mathkud, though, is unknown. Al-Idrisi, *La Première géographie de l'Occident*, 208.

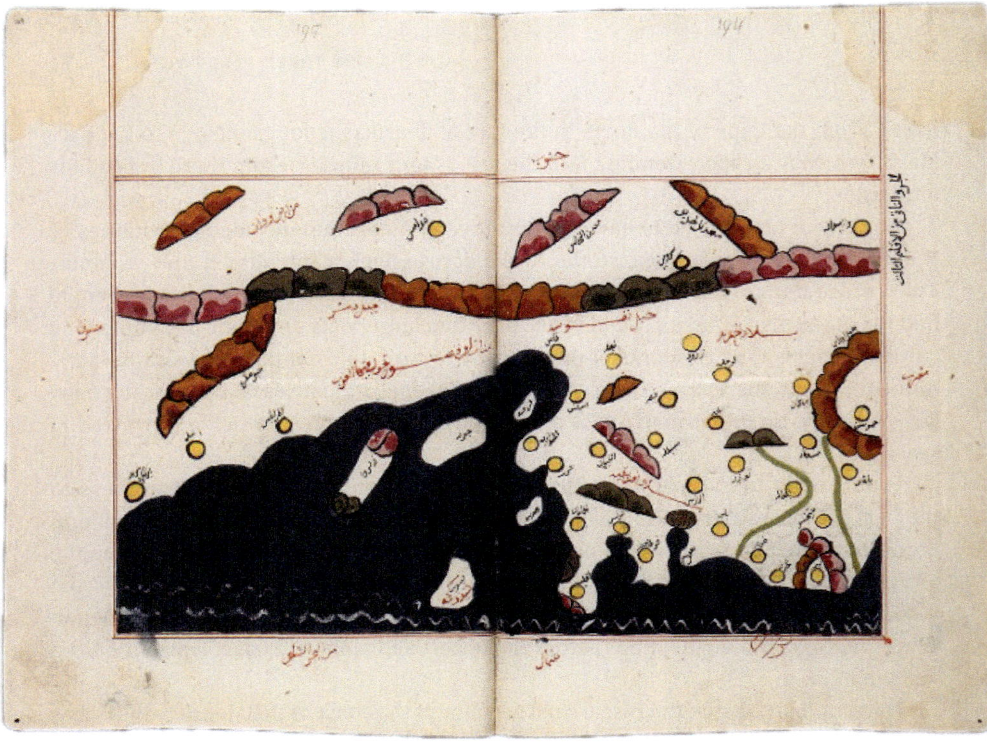

Figure 2. Map of Tunisia and Libya (Oxford, Bodleian Library, MS Pococke 375, fols. 102b– 103a).

Figure 3. Map of Morocco and Algeria (Oxford, Bodleian Library, MS Pococke 375, fols. 86b– 87a).

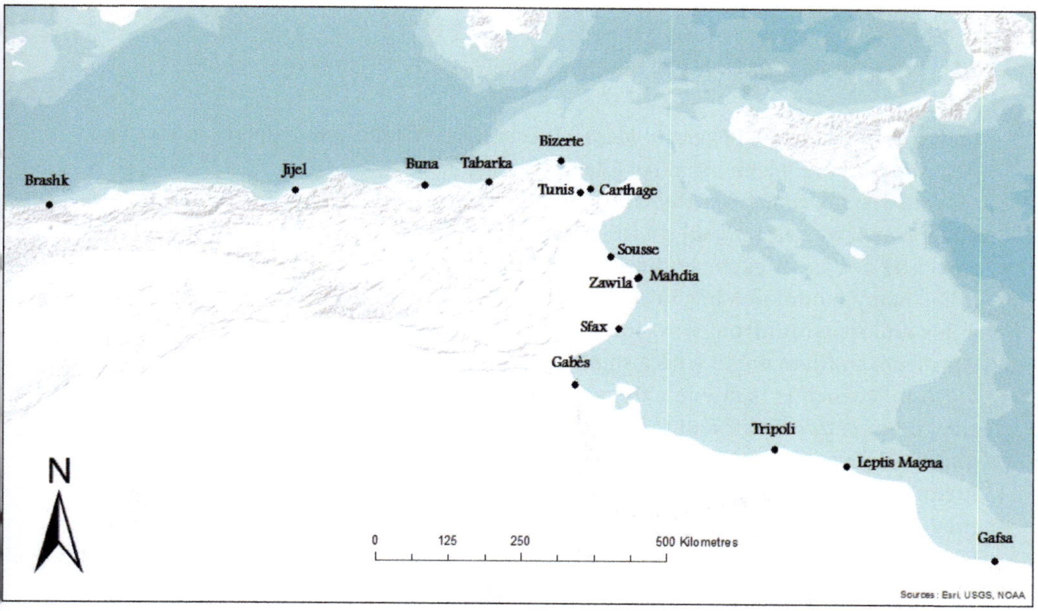


Map 1. Cities in Ifriqiya. Map by authors.

# CLIME FOUR, PART TWO

THIS SECOND PART of the Fourth Clime comprises a part of the Mediterranean including all the inhabited and famous as well as obscure and little-known islands and also a part of the inland countries of the Rum which we will now describe with Allah's aid.[1] We say that in this section amongst the large islands are Sardinia, Corsica and Sicily. There are also small islands such as Elba, Pianosa, Stromboli, Vulcanello [*Jabal al-Burkān*],[2] the islands of Vulcano [*Jazīrat al-Burkān*],[3] Lipari, Salina [*Dandima*], Asinara [*Um al-Ḥimār*],[4] *al-Ṭarfāniyya*,[5] Alicudi, Ustica, Giglio, Favignana [*Jazīrat al-Rahb*],[6] Pantelleria [*Qūṣra*],[7] Lampione [*Jazīrat al-Kitāb*],[8] Linosa, Comino, Malta and Marettimo. The coastal regions on the mainland are Barcelona, Girone, Ampurias, Narbonne and Carcassonne and these are all part of Gascony. In the eastern part of this section are

---

**1** Al-Idrisi is referring in a general sense to Christian territories including France and Spain.

**2** This is literally "The Mountain of the Volcano."

**3** This is literally "The Island of the Volcano."

**4** This is literally "The Mother of the Donkey."

**5** The location of this site is unknown. Amari and Schiaparelli suggest it could be one of the small islands located in the Strait of Boniface between Sardinia and Corsica. Al-Idrisi, *L'Italia descritta nel "Libro del Re Ruggero"*, 15.

**6** This is literally "The Island of the Monks."

**7** Al-Idrisi is using an Arabized version of the ancient name of Cossyra or Cossura.

**8** This is literally "The Island of the Book."

the lands of Calabria—Reggio, *al-Māṣa*,[9] Tuccio, Tropea and Santa Eufemia and we will describe each place with the aid of Allah.[10]

And we say that from the island of Manorca to the coast of Barcelona is one day's travel by sea [*majra*] and like this from Manorca to Sardinia is four day's travel by sea in an easterly direction. The island of Sardinia is a large territory, it has many mountains but little freshwater. Its length is two hundred and eighty miles and its width from west to east is one hundred and eighty miles. Its length runs from south to north with a small deviation to the east. There are three towns: *al-Qīṭna*[11] which is located in the south of the island, and which is a prosperous city; there is *Qālmara*[12] and this is located at the head of the straits between Sardinia and Corsica; the third city is *Qashīliyya*.[13] The people of Sardinia are of Latin African [*Rum Afārqa*] origin, the most barbaric and savage of the races of the Rum; they are brave and determined and they do not relinquish their weapons. On the island there is a good silver mine and from this mine silver is exported to many countries of the Rum. The length of the crossing between Sardinia and the island of Corsica is twenty miles.

The island of Corsica is made up of crags and gulfs and on its eastern side is the sea called, in the foreign tongue, *Ṭarāna* [Tyrrhenian]. In the western part of the island there is a pleasant city of moderate size which is prosperous. The island's length is one hundred and fifty miles, and its width is twenty-seven miles. Corsica is fertile and it has abundant cultivation. The Corsicans travel widely throughout the land of the Rum, and they are the most well-travelled Rum.

Amongst the islands that are adjacent to the land of the Rum is Elba. Between Elba and Corsica there is a distance of a day's travel by sea. Elba's circumference around its coast is one hundred miles and the island is part of the territory of Pisa. From Elba to the island of Pianosa to the north-east is twenty-five miles. The circumference of Pianosa is thirty miles, and the island is uninhabited. From Pianosa to the island of Capraia is thirteen miles and from there to the Pisa River is thirty miles. From Capraia to the island of Capri to the east is ___.[14]

Capri is an inhabited island, and its inhabitants are Amalfitan herdsmen. There is a medium-sized city and in the middle of the city there is a natural spring. Between

---

**9** This location is uncertain; it is possibly the village of Masa close to Ricadi.

**10** Bresc and Nef propose that *Tūjinsh* is Tuccio in the area of the Tuccio river valley. Al-Idrisi, *La Première géographie de l'Occident*, 301.

**11** This location is uncertain. Amari and Schiaparelli posit this could be the bishopric of Fausania (also known as Pausania/Phausania) established in the sixth century. Al-Idrisi, *L'Italia descritta nel "Libro del Re Ruggero"*, 16. See also Pier Giorgio Ignazio Spanu, *La Sardegna bizantina tra 6 e 7 secolo* (Oristano: Editrice S'Alvure, 1998), 97.

**12** This could be the modern site of Cagliari (on the southern coast of Sardinia), though the location does not make much sense in the context of al-Idrisi's description, which places it at the north of the island.

**13** The location of *Qashīliyya* is uncertain.

**14** Distance missing from all manuscripts.

Capri and the city of Sorrento in Calabria is twelve miles. In the east of the island there is a small port and between there and Naples is thirty miles. Between Capri and Ischia is sixty miles and the latter is a fertile though not very large island. It is close to the coastal city of Naples. On the island there are people of the Rum who live with their women and offspring in a pleasant city called *Mīyūr*. From this name comes the name of the island Ischia Maggiore.[15] Between Ischia and Naples is thirty miles, and between Ischia Maggiore and Ventotene is fifty miles. On Ventotene there is an excavated city, an arsenal and a port carved in stone; the sea enters through a canal which is excavated from the rock.[16]

Between Ventone and the city of Gaeta is twenty miles. Thirty miles in a south-westerly direction from Ventone there are two islands, one of them is Palmarola [*Mūnsa*] and the other is Ponza and they are both prosperous. From the island of Capri to the island of Stromboli in a south-easterly direction travelling close to the island of Sicily is ___.[17] Between Amalfi and Stromboli there is no other island except Capri. Stromboli is located to the north-east of the island of Vulcano, Stromboli has natural springs but does not have a port. The closest point on the mainland to Stromboli is Amantea in Calabria, and between the two is forty miles.[18] Between Stromboli and Vulcano is thirty miles.

The island of Vulcano is not large, but it has a large mountain which blazes some of the time with a great fire and during brief periods dies down. If this fire flares up, the mountain spews burning rocks and there is a terrifying roar which can be heard from a great distance away as if it were thunder. The island also has wild goats. Between Vulcano and the nearest point in Sicily is fifteen miles, and this is Tindari.[19]

From Vulcano in a westerly direction and a mile to the north is the island of Lipari, this island is populated only some of the time. Lipari has a castle and between Lipari and Vulcano it is four miles and on Lipari there is water, wood, and a small port. From the island of Lipari to the island of Salina is three miles north. This is a small island with no port. From the island of Lipari to the island of Filicudi is ten miles in the opposite direction from the qibla. This island is uninhabited, and there is no port.

---

**15** The location of *Mīyūr* is unknown. The term "Rum" here likely means Byzantine or Greek-speaking Christians. Ischia was under Byzantine control, albeit in a tenuous league with Naples, until Roger II took the island in 1130. Barbara Kreutz, *Before the Normans: Southern Italy in the Ninth and Tenth Centuries* (Philadelphia: University of Pennsylvania Press, 1991), 22.

**16** Ventone is probably the Villa Giulia, a Roman imperial villa used as a place of exile for women accused of adultery. Sarah T. Cohen, "Augustus, Julia and the Development of Exile 'Ad Insulam,'" *The Classical Quarterly* 58 (2008): 206–17.

**17** Distance missing from all manuscripts.

**18** Al-Idrisi's placement of Amantea does not align well with our modern geography, as Amantea is much further north than Stromboli. Amari and Schiaparelli claim that al-Idrisi may be confusing Amantea with Tropea. Al-Idrisi, *L'Italia descritta nel "Libro del Re Ruggero"*, 19.

**19** Al-Idrisi makes no mention of Tindari (ancient Tyndaris) in the description of Sicily that follows.

From Filicudi to Alicudi is ten miles, Alicudi is located to the south-east of Filicudi, the same direction as the *sirocco*.[20] This is a small island which offers refuge within its narrow port. It is forty miles between Alicudi and the island of Ustica. On Ustica there is freshwater and a safe port. Ustica faces Mount Pellegrino in the city of Palermo in Sicily and between the two is forty miles. To the south of the island of Ustica is the island of Favignana and in the southern and eastern part of the island there are ports where ships can safely anchor. There are freshwater wells, and the island is above Trapani, at a distance of fifteen miles. To the north of Favignana is the small island of Levanzo, but it has no water or port.[21]

The point on Sicily closest to Levanzo is the city of Trapani and between the two is ten miles. In a westerly direction from Favignana is the island of Marettimo which faces Tunis and Carthage and from Marettimo to the island of Favignana is thirty miles. There is no port on Favignana but there are goats and deer. To the east of Marettimo and to the south-east of Favignana is the island of Pantelleria. Pantelleria faces Nabeul in the territory of Ifriqiya and also faces the cities of Sciacca and Mazara in Sicily and between the two is one day's travel by sea and in the same way between Pantelleria and the mainland of Ifriqiya is one day's travel by sea.

The island of Pantelleria is small and fortified, it has wells, beaches, olive trees and there are many wild goats that were once domesticated. On the southern side of the island there is a secure port which is protected from various winds. In an easterly direction is the island of Gozo and between the two there is a distance of one hundred miles and on Gozo there is a secure port. From Gozo to the small island called Comino is ___.[22]

From Comino in an easterly direction is the island of Malta and this is a large island where there is a secure port which opens to the east. There is a city, and the island has plenty of pasture, sheep, fruits and honey. And between Malta and the nearest point in Sicily, a place called *Akranta* is eighty miles.[23] And beyond Malta in a westerly direction there is nothing until the island of Crete.

The island of Lampedusa is two days' journey by sea to the nearest point in Ifriqiya which is Kaboudia. On Lampedusa there is a secure port protected from all wind and it accommodates many fleets, this port is located in the southwest of the island.[24] There are no fruits or animals on Lampedusa. From Lampedusa in a northerly direction there is a lovely island called Lampione and between Lampedusa and Lampione is five miles, and this latter island is very lovely and is in a slightly westerly direction from

---

**20** The *sirocco* is a wind that comes off the Sahara and moves north-east over the Mediterranean, particularly in the summer months. Scirocco derives from the Arabic *shurūq* ("sunrise").

**21** Contrary to this statement, al-Idrisi writes in a later section that Trapani is near the islands of Favignana, Levanzo and Marettimo, each of which has a port, wells and wood.

**22** Distance missing from all MSS.

**23** Al-Idrisi does not mention this place in his description of Sicily. Bresc and Nef suggest this could be Capo delle Correnti. Al-Idrisi, *La Première géographie de l'Occident*, 305.

**24** Al-Idrisi writes "*fī al-libāj*," which corresponds to the modern *libeccio*, a south-western Mediterranean wind.

Lampedusa. From Lampione to Linosa in a north-easterly direction is thirty miles. Linosa has no port or woods but there is arable land. Attempting to anchor off Linosa is dangerous. The island of Gozo is located to the east of Linosa and is two days' journey by sea.

And we have provided important information about these islands without being too lengthy, many thanks to Allah. And what remains now is to describe the great island of Sicily. We will talk about its various regions, describe its localities, place by place, enumerate its glories and present its excellent qualities using in-depth research by the will of Allah almighty. And we say that Sicily is unique—the island is unparalleled in the beauty and remarkable qualities of its regions and inhabitants. From ancient times travellers have frequented the island, coming from far and wide. All these visitors agree on Sicily's superiority; they laud its worth, wonder at its splendour and pronounce its merits—the way the island combines within itself diverse attractions and all the bounties of the countries of the world. The dynasty of its kings is the most noble and their attacks against their various enemies are the most severe. Sicily's kings are the greatest of all kings, their greatness is terrible, their prestige is considerable, their loftiness is immeasurable and of the highest order.

It was in the year 453H[25] that he [Roger I] conquered the best of the territory [of Sicily], its tyrannical rulers and armies were vanquished by the greatest, the most virtuous and gallant, the most highly esteemed, the most praiseworthy, the honourable Roger, son of Tancred, the best of the kings of the Franks.[26] Roger did not rest until he had sent scurrying the island's prefects and the armies of the tyrants, mounting raids against them day and night, ultimately casting at them all kinds of death and ruin. Razor-sharp blades and the deadly lance were used until total victory was achieved—he conquered the island territory by territory and occupied each coastal city in turn and this all took place over a period of thirty years. Once Roger's will had been asserted and the stability of the throne ensured the path of justice was granted to the people of Sicily—he granted them the right to practise their religions and follow their laws; he ensured security in all things—the people's money, themselves, their families, and offspring.

Roger I lived out his days in this way until he could no longer put off the inevitable and while he was in Mileto in Calabria he died and was buried there in the year 494H. His son inherited the kingship after him, his namesake the great King Roger II. Roger II founded the state administration and embellished the magnificence of the realm, he brought honour to the sultanate, gave all matters their just share of his lucid perspective and good actions and through this disseminated justice, the establishment of security and worthiness to the kingdom. Owing to these actions, the various former rulers of these lands bent to Roger's will. They proclaimed their motto of allegiance to him, and followed him, handing the keys of their lands over to him. All sides came to be part of his kingdom and to come under his protection and

**25** 1043/44 CE.

**26** This is Tancred of Hauteville (ca. 980–1041).

mercy. His rule has only increased in prestige and nobility up to the point that we composed this book.

The afore-mentioned island of Sicily is of pre-eminent importance—its worth is significant, its territories vast, its regions many, its advantages abundant and its virtues impressive. And if we tried to enumerate all the island's virtues, and describe its conditions region by region, although this task is honourable, we would not have the adequate means. Therefore, we will have to summarise and express succinctly the island's many excellent features, so we say that:

Sicily, at the time of writing this book, is ruled by the great King Roger, and is made up of one hundred and thirty localities among them cities and castles, not counting estates [*ḍīyāʿ*], villages, and other places.[27] First, we want to describe the main coastal regions of the island, restricting ourselves to these parts until we are satisfied with their description and have returned whence we began. Then we will examine what is in the interior of the island, its regions, castles, and its extensive inhabited lands, from place to place and locality to locality, by the will of Allah almighty.

First is the city of Palermo, which is the great and beautiful city, the pulpit which stands above all the countries of the world. It is the proudest and most noble city and has been the seat of kings from the present to bygone days. It is the point from which troops and fleets have set off to conquer and the point to which they have returned, just as they still do nowadays. Palermo is situated on the coast, with the sea to the east and great, towering mountains encircling it. The city's coast sparkles with cheerful radiance. The city boasts magnificent buildings which travellers to the city wander around wondering at their excellence, their detailed features, and the originality of their design.

Palermo is divided into two parts. First, there is the Cassaro [*Qaṣr*] area, and second is the suburbs.[28] The Cassaro is the famous ancient palace renowned in each country and climatic zone; this area is made up of three zones.[29] Within the middle zone are found fortified palaces, imposing and noble dwellings, as well as numerous mosques, funduqs, bathhouses and the shops of prominent merchants. The remaining two zones also have gracious palaces and highly impressive buildings as well as numerous bathhouses and funduqs. The great mosque is to be found here; the mosque has returned to what it was in ancient times.[30] The mosque's appearance is difficult to

**27** Al-Idrisi uses the plural *ḍīyāʿ* (sing. *ḍīyʿah*), which means estates/hamlets/small townships. These *ḍīyāʿ* may have been lands belonging to the church of which al-Idrisi makes no specific mention in what follows.

**28** The Cassaro was the ancient, fortified nucleus (Punic then Romano-Byzantine) of Palermo that was the focal point of the city under the Aghlabids. Alessandra Bagnera, "The Urban Evolution of Islamic Palermo (9th–Mid-11th Century)," in *A Companion to Medieval Palermo*, ed. Annliese Nef (Leiden: Brill, 2013), 69.

**29** *Simāṭ* (pl. *Asmāṭ*) meaning 'row/series/order' also 'tablecloth/dressed table' on which a meal or banquet is served.

**30** Al-Idrisi is perhaps deliberately oblique here stating indirectly that the mosque was once a church and has now returned to being a church under the Normans.

comprehend due to the remarkable quality of its craftsmanship, the creative innovation of its varieties of pictures, embellishments, decoration, and inscriptions.

The suburbs are another city altogether and surround the Cassaro on all sides. This area includes the old city which is called Kalsa [*al-Khāliṣa*], and this is where the Sultan and the court once lived during the days of the Muslims.[31] The Door of the Sea [*Bab al-Bāḥr*] is in Kalsa as well as the arsenal where ships are constructed.[32] Waters surround and flow throughout the city of Sicily and its springs are gushing. The city's fruits are abundant and its buildings and parks so lovely that they dazzle those who try to describe it, mesmerising the mind; the sight of it, in short, seduces all those who behold it.

The castle previously mentioned is one of the most indestructible fortresses; it is elevated and impervious to the damage wrought by warfare, it is impregnable. At the top of the castle there is a fort recently added by the great King Roger and built using small pieces of mosaic and huge blocks of carved rock. The palace has been fortified in keeping with its artistic arrangement; its towers and bulwarks have been reinforced. The king's palaces and assembly halls have been perfected, and the magnificence of their construction and design, their unique details, are praised by foreigners and travellers who declare that there is no greater city in the world for the graciousness of its villas and palaces.

The suburbs surround the castle that we have previously mentioned; this is a large area with numerous dwellings, funduqs, bathhouses, shops, and markets. There is a wall that surrounds the district with a moat and a trench. Within the district, there are many orchards, wondrous parks and running streams of freshwater which come down from the surrounding mountains. Beyond the suburbs to the south is the Oreto [*ʿAbbās*] river and this is a running river upon which there are mills that are sufficient to meet the needs of the city.

In an easterly direction from the city, a day's journey away, is the fortress of Termini situated on a hillock overlooking the sea. This is one of the most exalted fortresses and its location is amongst the largest. There is a wall that surrounds the fortress and there are original monuments and ancient buildings including a remarkably constructed amphitheatre which demonstrates the construction abilities of its creators. There is also a new castle and two lovely hot springs constructed side by side with an ancient building upon them.

To the west of Termini is a place known as Trabia [*al-Tarbīʿa*] and this is one of the loveliest places.[33] There is running water, numerous mills and wide tracts of culti-

---

**31** *Al-Khāliṣa* is literally "the Chosen One/the Pure/the Elect." The Kalsa was constructed in the tenth century as a fortified area opposite the Cassaro to protect and separate the Shi'ite Fatimid governor, elite bureaucrats, and the military from the primarily Sunni urban population. Ferdinando Maurici, *Palermo araba: Una sintesi dell'evoluzione urbanistica (831-1072)* (Palermo: Kalós, 2015), 70.

**32** Many coastal cities in the Islamicate world had a *bab al-bāḥr*, including Málaga, Almería, and Tunis.

**33** *Al-Tarbīʿa* is literally "The Square/Squareness."

vated and grazing land. Pasta is produced here which is exported in large quantities to all regions, including all parts of Calabria, as well as to the Muslim and Christian lands and many loads are carried away. In Trabia is the Termini River [*al-Silla*], this river is large and has an abundance of water. In springtime, the fish known as *Ray* is caught here, and at the river's port the large fish known as tuna is fished.

Twelve miles away is the castle of Brucato, this is a towering castle which has much agricultural cultivation, a market, various facilities, abundant water supplies, many mills, orchards, gardens, vast estates, and verdant crops. The sea is two miles from Brucato. From Brucato to Campofelice di Roccella [*Ṣakhrat al-Ḥadīd*] is twelve miles.[34] This is a small estate with a castle that sits atop the rock already mentioned, the rock descending precipitously towards the sea. On the landward side there is a level stretch of sand with fertile farmland and verdant crops.

From Campofelice di Roccella to Cefalù is an easy day's journey. The castle of Cefalù is on the coast and resembles a city with markets as well as a bathhouse and mill located on a spring which flows into the city centre, the inhabitants of the castle drink this water which is fresh and cool. Cefalù is built on rocks on the edge of the sea and there is a good port which is visited by ships from every region. It is a well-populated area and there is a fortress overlooking the town at the top of an imposing mountain which is hardly ever surmounted due to the difficulty of the ascent.

From here to the castle of Tusa is an easy day's journey and this is an ancient, impregnable castle. Attached to the castle is a prosperous quarter, both of which are situated at the top of an isolated mountain which is very hard to reach as you need to traverse rough terrain and difficult routes to reach them. Tusa is surrounded by land with fertile soil, wide tracts of land, excellent crops and other advantages. Tusa is about two miles from the sea. From Tusa to the Fortress of the Boats [*Qalʿat al-Qawārib*] is twelve miles.[35] This is an elevated and very old fortress, of ancient construction, encircled by a prosperous village. Its farms are verdant, its crops abundant and water copious. There is a well-frequented port where ships can dock and carry loads away and between the fortress and the port there is a distance of about a mile and a half.

---

**34** *Ṣakhrat Al-Ḥadīd* is literally "The Rock of Iron." Al-Idrisi is referring to the Castello di Roccella the ruins of which are situated in the modern town of Campofelice di Roccella. Maurici, *Castelli medievali in Sicilia*, 357.

**35** The location of *Qalʿat al-Qawārib* is uncertain. Amari identified this place as Santo Stefano Vecchio (also known as Santo Stefano di Mistretta), which was destroyed by a landslide in 1682 (the settlement was subsequently moved to the coast and was renamed Santo Stefano di Camastra). Santo Stefano Vecchio is located in the area of Contrada Lavanca, where excavations led by the Comune di S. Stefano di Camastra have been underway for over a decade. However, there is no ancient fort located here. Sicilian archeologists involved in the excavations at Contrada Lavanca have convincingly argued that al-Idrisi may be in fact referring to the castle of Mistretta and surrounding village (ancient Amestratos). Nuccio Lo Castro, ed., *Santo Stefano di Camastra: La città del duca* (Messina: Comune di S. Stefano di Camastra, 2012), 23–25; Amari, *Biblioteca Arabo-Sicula, Versione Italiana*, 1:65.

From the Fortress of the Boats to Caronia is twelve miles. Caronia is the first town in the province of Val Demone [*Damunash*].[36] It is an ancient fortress, but it has a new fort. Caronia has gardens, rivers, vineyards, trees, and a port on the coast; from the port a net is used to catch a great amount of tuna. Between the fortress and the sea there is a distance of about a mile. From Caronia to San Marco d'Alunzio is ten miles and this is a grand fortress with ancient monuments and extensive buildings. There are markets, a bathhouse, and all varieties of fruit. There is grazing land, plentiful crops, freshwater, and vast territory. Violets grow all around the town and their scent perfumes the air, there is also plentiful silk. San Marco's coast is beautiful, and boats are built there using wood from the surrounding mountains.

From San Marco d'Alunzio to the castle of Naso is ten miles, Naso is elevated and its territory vast. The area is prosperous, and its rivers are gushing, it has gardens and streams. There are cultivated fields and mills. Its coast is delightful and its castle in a pleasant position and between Naso and the coast there is a distance of two miles. From Naso to Patti is twelve miles. Patti is an impregnable castle, and its territory is expansive. It has thriving fields, charming dwellings, flowing water, many gardens, and sublime countryside. Patti overlooks the sea at a distance of a mile. From Patti to Oliveri is three miles. This is a delightful estate with an imposing fort situated on the coast. Oliveri has a market and bathhouse and is well-populated. In addition, there are vineyards, running water, cultivated fields and mills, there is also a good port where much tuna is caught.

From Oliveri to the castle of Milazzo is twelve miles and this is a large fortress situated on a cape which juts out into the sea. The attractiveness of its appearance, its solid construction, its sublime countryside, and its impregnable fort make it one of the best places of all, the most beautiful and most elegant. Milazzo resembles a great metropolis in terms of its buildings, its organization and markets, its commodities, and amenities. Milazzo is on the coast and the sea surrounds the town on all sides, except its northern side which is the point from which you can enter to it. It is possible to travel there by land or sea. Linen of excellent quality is exported from the town, there are fertile fields, gushing water and traps which catch abundant tuna. From Milazzo to Messina is an easy day's journey.

The city of Messina is at the eastern corner of the island and mountains surround the city on its western side. Its coast is delightful and its land fertile; there are gardens and orchards with abundant fruits, there are gushing rivers with numerous mills upon them. It is indeed one of the most splendid regions, the most prosperous, and travel to and from the city is constant. The arsenal and the main dock for all the maritime states of the Rum are here; this dock is the gathering point of huge ships as well as travellers and traders from all Christian and Islamic countries. Visitors come to Messina from every region, its markets are splendid, their merchandise in high demand. In the mountains there is an iron mine, and this iron is exported to neighbouring regions. The port is one of the great wonders, spoken about throughout all the countries; the

**36** *Damunash* is the ancient name for the Val Demone, one of the three administrative zones of the island under Muslim rule, along with the Val di Mazara and the Val di Noto.

biggest ships must dock on the beach and from there their cargo is unloaded by hand. Located at Messina are the straits which take you to Calabria; this stretch of sea is difficult especially when the wind drives against the sea's current. When the sea enters the straits from both sides at the same time, nothing could save you from disaster save the grace of Allah most Exalted. At its widest point the straits are ten miles in width and at its narrowest, three miles.

From Messina along the coast to Taormina is a day's journey. Taormina is an impregnable castle located at a precipitous and lofty position. It is one of the main ancient fortresses of the island and one of the most revered age-old regions. Taormina is on a mountain overlooking the sea, there is a good port which receives visitors from far and wide and many crops are carried away. There are dwellings and markets, and it is the meeting point of caravans and companies before they proceed to Messina. There are excellent estates, fertile agricultural fields, a gold mine, and the famous mountain *al-Ṭūr* known in verses as being a place of worship.[37] There are rushing rivers, many mills, and a small number of gardens. There is a river over which there is an amazing bridge and its construction hints at the skills of its builders and the power of the Sultan. There is an amphitheatre, which is one of the oldest amphitheatres of the Rum and also represents the sublime qualities and prestige of the king. There is a gold mine. From here to Aci Castello is one day's journey.

Aci Castello is a town on the seaside, and it is one of the most ancient. It has a market, grazing land, and fertile fields; the hot Clime means that crops are harvested here before anywhere else on the island. Pitch, tar, and wood as well as many other products are exported from Aci Castello. To the west of Aci Castello is the mountain known as Mount Etna [*Jabal al-Nār*].[38]

From Aci Castello to the city of Catania is six miles. This is a beautiful region, known as the 'region of the elephant'; it is a place of prestige and renown. Catania is on the coast and there are abundant markets, splendid dwellings, mosques both large and small, bathhouses and lodgings [*khānāt*].[39] There is a good port which is visited from far and wide, and all manner of merchandise is carried away. There are many gardens and Catania's water supply comes from its gushing rivers and springs. There is a river which presents a curious and marvellous phenomenon—in some years it floods dramatically, branches off into streams and mills are concentrated along its banks; in other years, the river completely dries up and you cannot even find enough water for a drink. Catania's agricultural territory is abundant, its grazing land and farms both fertile and productive, the city walls are impenetrable, and the surrounding territory is vast. The elephant for which the town is famous is a stone talisman in the shape of

---

**37** Al-Idrisi is likely referring to Monte Tauro, the site of the ancient city of Tauromenium that overlooks the Greco-Roman theatre. Luigi Santagati, *La Sicilia di Al-Idrisi ne "Il libro di Ruggero"* (Palermo: Sciascia, 2010), 52.

**38** This is literally "The Mountain of Fire."

**39** Sing. *khān*. There were two types of *khan*. One was a staging-post along a commercial route; the other was a kind of depot in an urban setting where merchants could await the time for the transfer of merchandise and procure lodging. See "Khān," *EI2*.

an elephant, in the past it was erected on top of a building but now it has been brought into the city centre and has been erected in the church of the monks. To the west of Catania is the Simeto [*Mūsa*] river, the great river which flows into the sea.[40] There are many fish throughout the river of extreme size and excellent taste. The cities of Taormina, Aci Castello and Catania lie at the eastern foot of Mount Etna which we have already mentioned.

From Catania to the fortress of Lentini is a day's journey. Lentini is a strong and civilized fortress and has markets much like a city. Lentini is six miles from the sea; it is located on the bank of its eponymous river. Boats travel up the river with their cargo until they arrive opposite Lentini on its eastern side. To the west of Lentini the countryside is vast and widely spread out. In the river there is a great variety of excellent fish unlike any other, they are exported far and wide. In Lentini there are abundant markets, funduqs and a large population and from Lentini to Syracuse is a long day's journey.

Syracuse is one of the most famous and notable cities and attracts city dwellers and peasants alike; merchants come to the city from all regions. The city is on the coast and the sea encircles it from every direction; to enter or leave Syracuse you must do so through a single gate on the city's northern side. Syracuse's praises have been so widely sung it seems unnecessary to describe the famous locale and well-known stronghold any further. There are two ports the like of which cannot be found elsewhere in the world, one is larger than the other and this one is to the south, the other one is more famous and is to the north. The fountain of Aretusa [*al-Nabūdi*] stems from a cliff on the edge of the sea and it is wondrous. In Syracuse you find the following, which are all on par with the largest cities: cloth markets, khans, houses, bathhouses, spectacular buildings and vast squares. Syracuse has a vast territory, with estates and dwellings as well as fertile land and good farms. Ships load up with food and other products for export to all countries and regions; the city has gardens which produce fruit which exceeds all limits. From Syracuse to Noto is a day's journey.

Noto is one of the most secure fortresses and one of the most beautiful and noblest cities. Its territory is vast, and it has many excellent benefits; it has lovely, well-ordered markets and meticulously constructed dwellings. Its rivers are freely flowing, gushing with water and there are many mills. Noto is eight miles from the sea. Noto has widespread agricultural fields; its farms are the most fertile and its lands are the most abundant. Noto is ancient and contains age-old monuments. From Noto to the sea is eight miles.[41] Between Noto and the sea is the pleasant estate Cassibile where farms are spread over a wide area. From Noto to the eastern edge of the island is one day's journey. This area is completely deserted, and the corner section is called Porto di Ulisse.

From Noto travelling along the coast is a day's journey to Scicli. Scicli is a fortress atop a mountain, one of the greatest fortresses in one of the best positions. It is about

**40** *Mūsa* is Arabic for "Moses."

**41** It is not clear why al-Idrisi repeats the distance between Noto and the sea.

three miles from the sea. It is a noble place; its people are prosperous, and it is a civilized place. There are markets with goods that come from all over. Scicli has abundant blessings and vast lands with gardens that produce every type of fruit. People travel by sea to reach Scicli from every part of Calabria, Ifriqiya, Malta and other areas. Its plains are the most fertile, its farms the most invaluable, its territory vast and fecund with perfect conditions; there is a gushing river with many mills upon it. There is the spring known as the Spring of Hours [*ʿAin al-Awqāt*] that demonstrates a strange phenomenon: the spring produces water in the hour of prayers and ceases at other times.

From Scicli to Ragusa is thirteen miles. This is an impregnable fortress and a noble and ancient settlement. Rivers and streams encircle the town and there are many mills, delightful buildings, and vast squares. There is a fertile plain with abundant farms. Between Ragusa and the sea there is a distance of seven miles. Ragusa's river, from which the town takes its name, flows past on its eastern side and at the mouth of the river on the sea there is a good port where boats come to load and unload. There is a market where people come to trade from far and wide. Between Ragusa and Butera is two day's easy journey—it is forty-five miles between the two places.

Butera is an impregnable and renowned castle, amongst the best places whether urban or rural, it is similar to large and prosperous cities. Its buildings are beautiful and well-constructed, its dwellings wondrous, its markets large and of the best standard. There are mosques for congregational prayer, a bathhouse and khans. One of the greatest rivers encircles the town and there are gardens on every side with many excellent fruits and many wondrous blessings. Between the town and the sea there is a distance of about seven miles. Between Butera and Licata is one day's journey which is a distance of twenty-five miles.

Licata is a castle on top of a rock and the sea and river surround it. You cannot enter the place except for one gate at its northern end. There is a port at which boats arrive, taking away loads of merchandise. Licata is prosperous and has a market, its lands are vast, and it has fertile farms. The mouth of its river at the sea is called the Salso [*al-Wādi al-Milḥ*] and within its waters there are many plump, tasty fish that are delicious to eat.[42] From Licata to Agrigento is one day's journey which is twenty-five miles.

Agrigento is one of the most civilized cities, frequented by people who come and go. Its fortress is supreme and impregnable, it is a beautiful and ancient city, famous in all countries. It is one of the most secure fortresses and one of the most glorious territories. People are drawn here from far and wide, and it is the meeting place of travellers and ships. Its dwellings are amongst the most splendid of all dwellings, they captivate the beholder. There are markets with all manner of goods to trade and buy; there are gorgeous parks and gardens where many types of fruit are available. The city's ancient monuments attest to its grand past. The great ships that come to Agrigento exceed their capacity within a few days due to the abundance of what is available. There are gardens and well-renowned produce, and the city is three miles from

**42** This is literally "The Salt River."

the sea. From Agrigento to Sciacca is one day's journey along the coast which is a distance of twenty-five miles.

Sciacca is a town situated on the coast; its position is open and elevated. Sciacca is prosperous with many markets and dwellings. At this time, it is the capital of the regions that surround it. Its port is constantly busy and travel here from Ifriqiya and Tripoli is always considerable. Sciacca and Caltabellotta [*Qalʿat al-Balūṭ*] share the same territory.[43] Caltabellotta is an impregnable castle, it is in a lofty position and difficult to reach. It has excellent grazing land, fertile farms and unusual types of fruits. There are springs and rivers with many mills upon them. In the past there were many people in the town but most of them have now moved to Sciacca; there remain only a few men to guard the fortress from those who wish to take it. It is twelve miles from here to the sea and it is nine miles to Sciacca. From Caltabellotta to Agrigento is a long day's journey and to Mazara is two days' light journey. Between here and Mazara along the coast is a large estate known as the Idols [*al-Aṣnām*].[44]

Mazara is an exalted city and there is no place to rival it in terms of prestige, the beauty of its appearance and construction and its combination of merits not found in any other city. It has towering, fortified city walls, delightful dwellings, a wide alley, streets, bustling markets with many types of trades and products, delightful bathhouses, extensive shops and fertile orchards and gardens. Mazara is frequented from far and wide by merchants who take away an abundance of goods for export. Its territory is capacious and includes many pleasant estates. A river known as the Mazara [*Majnūn*] runs along the city's walls and large boats travel along this river to load up with goods and smaller boats dock for the winter.[45] Between Mazara and Marsala [*Marsa ʿAli*] there is a distance of eighteen miles.[46]

Marsala was an ancient city, one of the most noble of Sicily but it was attacked and obliterated. Count Roger I surrounded the city with a strong wall which brought back prosperity, trade, and the collection of taxes. Marsala has a wide and extensive territory and people from Ifriqiya travel to the city frequently. Marsala's population drinks from freshwater wells that can be accessed between their houses and from springs that surround the city. Marsala has funduqs, bathhouses, orchards, and verdant farms. From Marsala to Trapani is a day's journey which is a distance of twenty-three miles.

Trapani is an ancient city situated on the coast and the sea surrounds the town from every direction. You reach the city via a bridge at the eastern gate and the city's port is in the south of the city. It is a calm port; most ships dock there for the winter, safe from all conditions and the raging seas. Fish of the finest quality are caught here, and the large fish known as tuna is also caught with big nets. The finest coral is harvested in its seas and at the city gate are salt marshes. The territory of Trapani is vast and extensive, and its lands are one of the most abundant agricultural areas in terms

**43** This is literally "The Fort of Oak."

**44** This is literally "The Idols," probably in reference to the Temple of Hera at Selinunte.

**45** This is literally "The Crazy River."

**46** Marsala is derived from the Arabic *Marsa ʿAli*, which literally means "The Port of ʿAli."

of quality and yield. Trapani has large markets and abundant sustenance. Near to Trapani are the islands of Favignana, Levanzo and Marettimo; each of these islands has a port, wells, and wood. People travel to Trapani in winter due to the quality of its port and the mildness of its sea and winds at this time.

From Trapani to Erice [*Jabal Ḥāmid*] is about ten miles.[47] This is a huge, imposing mountain easy to defend due to the difficulty of its ascent. At the top of the mountain there is level land which is good for agriculture, with plenty of water and there is a castle without guards watching over it. From Erice to Calathamet [*al-Ḥamma*] is a distance of twenty miles.

Calathamet is a towering, impregnable fortress, well-known as one of the finest fortresses. The sea is about three miles to its north. There is a port which has a fort built upon it known as the Stairs [*al-Madārij*].[48] Boats come and go from the port and tuna is fished using nets. This fortress was named the Geyser [*al-Ḥamma*] because of a hot spring whose waters flow out of a cliff nearby.[49] People bathe in this spring; its waters are fresh and of moderate temperature. There are rivers and streams nearby which have mills upon them. There are orchards, gardens, buildings and parks as well as lots of fruit. There are vast tracts of land and excellent land for agriculture. From here to Trapani is an easy day's journey.

From Calathamet to Calatubo is ten miles. This is an impregnable castle with a vast area of agricultural land and farms which are good for growing, this is an area with many advantages. Between Calatubo and the sea is a distance of about four miles; there is a port and boats come to load up with a great amount of food including different types of grains. There is a mine where two types of millstones are cut—the water stone and the Persian millstone [*al-fārsiya*]. From Calatubo to Calathamet is ten miles and from Calatubo to Partinico is twelve miles.

Partinico is a beautiful region, with a pleasing aspect. There is thriving agricultural land where much cotton is cultivated, and henna as well as different types of legumes. Water is plentiful and there are many mills. The fortress which is called Partinico is in an area called the Cemetary [*al-Jabān*] overlooking the town. There is a port about two miles north of the city called the Corner [*al-Rukn*]. From Partinico to Cinisi ___.[50] Cinisi is a vast estate, situated on the side of a mountain that overlooks the town. It has a wide tract of land, well-suited for growing crops, with excellent pastures and abundant fruit. The sea is to the north about four or so miles away.

From Cinisi to Carini is eight miles. This is a beautiful and pleasant region which is very secure. There are many types of fruits here and there is a large market as well as other features of civilized places such as markets, bathhouses and large dwellings. Carini produces a lot of almonds, dried figs and carob. Large and small boats load up with these goods, transporting them to a variety of places. Water is plentiful, available

---

**47** This is literally "The Mountain of Hamid."

**48** Castellamare del Golfo.

**49** The geyser likely refers to the hot springs at Terme Segestane.

**50** Distance missing from all manuscripts.

all over the town, the bulk of it flowing through to the city's gardens. There is a new castle that sits on a hill overlooking the town. The sea is to the north, about a mile away and between here and the great city known as Palermo is twelve miles.

These are the thirty-five regions of note on the coast. Together with these are the interior settlements which are numerous and are made up of castles and fortresses and other settlements and areas. We will mention all these, fortress by fortress, and castle by castle, by the will of Allah.

Firstly, we will begin with the area between outside Palermo to Enna [*Qaṣryānī*] in the middle of the island.[51] From the Palermo to Misilmeri [*Manzil al-Amīr*] to the east is six miles.[52] Misilimeri is a great and impregnable castle with plentiful water, territory and agricultural land. From Misilimeri to the Reservoir [*al-Khazzān*] is six miles.[53] This is a castle at the summit of a mountain—one of the most beautiful castles, situated in one of the best spots with the best conditions. It is prosperous and has estates. From the Reservoir the river called Eleuterio [*Wādi al-Amīr*] emerges, then descends along various channels and meets up with the waters of *Qajāna*, with *Qajāna* located on the northern side of the river.[54] Between *Qajāna* and *Jifla* is nine miles.[55] The river then arrives below Marineo with Marineo on the right side of the river and between Marineo and *Qajāna* there is a distance of one and a half miles. The river then flows below Misilmeri with the town on the river's northern side and between the two there is a distance of one mile. Between Marineo and Misilmeri there is a distance of six miles and from Misilmeri to the sea is one 'big' mile. From the Reservoir to *Jifla* is half a day's journey—it is about ten miles. Likewise, from Misilimeri to *Jifla* it is the same distance but is a day's journey.

*Jifla* is a pleasant region, a vast territory with agricultural fields and estates. Its waters and streams are gushing and plentiful. Its farmland and the confines of the region are vast. From the Reservoir to Vicari is fifteen miles. Vicari is an elevated and well-defended castle. There is flowing water and lots of fertile cultivated land, it is one mile from here to the Termini River, which descends into Termini. Vicari has continuous farmland; it has abundant cultivation and myriad blessings. From Vicari to

---

**51** *Qaṣryānī* means "The Castle of Yānī." The Italianized name Castrogiovanni was used for the city until 1926.

**52** This is literally "The Estate of the Prince."

**53** This location is uncertain. Amari places it near the palace of La Ficuzza in what is now the Bosco della Ficuzza Nature Reserve, while Bresc and Nef place it nearby at the highest peak in Western Sicily, the Rocca Busambra. Amari, *Biblioteca Arabo-Sicula*, 1:84; al-Idrisi, *La Première géographie de l'Occident*, 321.

**54** *Wādi al-Amīr* means "The River of the Prince." The location of *Qajāna* is uncertain. Santagati identifies this possibly as a farmstead at Rocca Fiori—about five kilometres west of Marineo. Santagati, *La Sicilia di Al-Idrisi*, 91.

**55** Amari places *Jifla* within the "comune di Cefalà," while Bresc and Nef identify it as the town of Gangi near Caltanisetta (though this placement is difficult to square with the distances provided by al-Idrisi). Amari, *Biblioteca Arabo-Sicula*, 1:85; al-Idrisi, *La Première géographie de l'Occident*, 321.

Petterana is nine miles.[56] Petterana is an impregnable and inaccessible castle. It has farms and crops, and its lands connect with those of Vicari, mentioned previously.

From the Reservoir to Jato is about fifteen miles. The castle of Jato is elevated and incredibly secure. It is the epitome of fertility and has a vast territory. There is an underground prison and those who attract the wrath of the king are committed there. There is no running water and there are no rivers nearby. From Jato to Calatrasi is nine miles.[57] Calatrasi is a renowned and ancient castle of very solid construction. It has farmland and the northern border of its lands connects with Jato and in the south with the castle of Corleone; between Calatrasi and Corleone is a distance of about eight miles. Between Corelone and *Qalʿat at-Ṭarīq* to the north is nine Arabic miles, which is three Frankish miles.[58]

Corleone is an impregnable castle, very well-fortified. It has extensive farmland and a river which is named after the town runs through it. From Corleone to *Rāya* is eight Arabic miles and between Corleone and Jato is five Frankish miles.[59] Between Corleone and Prizzi to the east is ten miles. Prizzi is a pleasantly situated castle which is very secure. It has a well-populated village, running water, springs, extensive and bountiful farms, abundant and lucrative blessings. From Prizzi to Castronovo is about twelve miles and from Castronovo to Corleone is twenty miles. Castronovo is a pleasant place replete with advantages and blessings, it has farmland, crops and running water. From Castronovo westward to *Rāya* is about ten miles and from Prizzi to *Rāya* is ten miles and from Corleone to *Rāya* is eight miles. Prizzi is in the north, Castronovo is in the east, Corleone is in the west and *Rāya* is in the south. *Rāya* is a noble estate, of excellent quality, there is rich and bounteous farmland.

The Termini River emerges on the western side of the foot of this mountain called *Rāya*. The river encircles the mountain then runs from the north where it meets the waters of Prizzi three miles to the right of the town in an easterly direction. The river then reaches the Màrgana estate which sits to the north of the river and between Màrgana and the river is one mile. Between the castle of Prizzi and Màrgana is four miles. The river then runs on the right-hand side under Vicari, and it is one mile between Vicari and the river. Between Màrgana and Vicari is three miles.

Here, the Termini River meets the San Leonardo [*Rīghnū*] River whose course begins at the *Zarāra* mountain at a place called Godrano [*al-Ghudrān*].[60] The waters of Mezzojuso [*Manzil Yūsuf*] are then added to the river which continues on with

---

**56** Known also as *Pitirrana* and *Peterrana*, the specific location of this site is modern Pizzo Pipitone. Maurici, *Castelli medievali in Sicilia*, 347.

**57** Maurici places Calatrasi at Monte Maranfusa. Maurici, *Castelli medievali in Sicilia*, 268.

**58** *Qalʿat at-Ṭarīq* means "The Fortress of the Road." Its location is uncertain, though Santagati speculates that it might be the ruined fort at Casale di Sopra in the area of Rocca Busambra. Santagati, *La Sicilia di Al-Idrisi*, 96.

**59** Literally "A banner/flag." This location is uncertain. Bresc and Nef state it is within the Commune di Palazzo Adriano, while Santagati places it below the castle of Calatialci and Pizzo Cangialoso. Al-Idrisi, *La Première géographie de l'Occident*, 323; Santagati, *La Sicilia di Al-Idrisi*, 97.

**60** The location of *Zarāra* is unknown. *Al-Ghudrān* means "The Streams" and corresponded to the

Mezzojuso on its right side, it then flows into the river under Vicari.[61] The river then extends to Petterana with Petterana on its right side and between Petterana and the river is three miles and between Vicari and Petterana is nine miles. Then the river flows to *al-Abrajā* with *al-Abrajā* on the river's right side and between the two is a distance of three miles.[62] Between *al-Abrajā* and Petterana is two miles. The river then flows to Caccamo with Caccamo on the river's right side and between Caccamo and the river is two miles and between *al-Abrajā* and Caccamo is one mile. The river then flows to Termini with Termini to the right and between Caccamo and Termini is ten miles, at this point the river flows into the sea. Between *Jifla*—which we have already mentioned—and *Khāṣwā* is two Frankish miles.[63] Furthermore, between *Khāṣwā* and Vicari is two Frankish miles. *Khāṣwā* is an estate with plentiful cultivation, a variety of blessings, grains and crops.

From Corleone to Battalari in the south is four Frankish miles. Battalari is an ancient castle of secure construction, mountains surround it and there is plenty of water. From Battalari to Caltabellotta—which we previously mentioned—is ten miles and from this fortress to Sciacca is four Frankish miles, this is twelve miles. From Calatrasi to the Estate of the Woman [*Raḥl al-Mar'a*] is eighteen Arabic miles.[64] This estate is prosperous, with plentiful farmland, fertility, dairy products and butter. From the Estate of the Woman to Partinico is one day's easy journey, it is about eighteen miles. To the west of the Estate of the Woman, along the Mazara Road, is the Idol [*al-Ṣanam*] after nine Arabic miles.[65]

The Idol is a large estate comprising many people, it has a towering, secure castle in a premium position. There are rows of trees, it is surrounded by orchards and water is plentiful. From The Idol to Mazara is about seven Frankish miles, we have already mentioned Mazara as it is a big city. Between Mazara and the Idols is three Frankish miles. We return now to Castronovo which we have already mentioned and from which the Platani River rises, gushing with water. The river then flows to Cammarata, then to Platani, and then to the sea.[66] From Castronovo to Cammarata is ten miles. From Cammarata to Platani is thirty miles, or one day's journey. Cammarata is a large

---

ancient site of Godrano, which is less than two kilometers from the modern town. Santagati, *La Sicilia di Al-Idrisi*, 101.

**61** *Manzil Yūsuf* means "The Estate of Yusuf."

**62** The modern location of *al-Abrajā* is uncertain. Bresc and Nef identify it as Libricci, while Santagati places it at Cozzo Casale. Al-Idrisi, *La Première géographie de l'Occident*, 323; Santagati, *La Sicilia di Al-Idrisi*, 100.

**63** The site of *Khāṣwā* is uncertain. Bresc and Nef, as well as Santagati place it at Pizzo di Casa, near Mezzojuso. Al-Idrisi, *La Première géographie de l'Occident*, 324; Santagati, *La Sicilia di Al-Idrisi*, 102.

**64** Santagati places this estate (with some reservation) at Finistrelle—some four kilometres west of Gibellina. Santagati, *La Sicilia di Al-Idrisi*, 102.

**65** Not to be confused with "The Idols," see note 44. According to Santagati, this site is modern Salemi. Santagati, *La Sicilia di Al-Idrisi*, 104.

**66** Maurici places Platani at Monte della Giudecca. Maurici, *Castelli medievali in Sicilia*, 350. Al-Idrisi describes Platani twice.

estate, with a wide expanse of farmland and much cultivation. It has a very secure castle located in an elevated position. There are orchards, gardens, fruits and blessings. In similar fashion, the castle of Platani sits in an imposing position, its fortress is splendid and rises upwards to a peak. Between Platani and the sea is about six miles.

Returning to the castles of Jato and Calatubo, both of which we have already mentioned, and between the two is five Frankish miles. From Calatubo to Alcamo is one and a half Arabic miles. Alcamo is a large estate, there are farms and fertility; it has a permanent market with artisans and industry. *Mīrjā* is one mile northward.[67] It is a small, secure castle, and it has a well-populated suburb as well as fertile land. From *Mīrjā* to Calathamet, which we have already mentioned, is one Frankish mile and from Calathamet to the Stairs is two Frankish miles.

The castle of the Stairs is the most secure of all castles in terms of construction and position; it has a trench encircling it which is cut into the mountain. You gain access to the castle by crossing a wooden bridge, which can be drawn up or down whenever desired. There are orchards and vineyards, and various fruits. There is a narrow port. From the Stairs to Calatubo is three Frankish miles, we have mentioned this place previously. From Calatubo to Partinico is three Frankish miles, we have already mentioned Partinico. From Partinico to Jato is eighteen miles and we have already mentioned this place too.

Returning now, we say that from Calathamet to Calatafimi is about eight miles. Calatafimi is an ancient castle, a stronghold of excellent construction. It has a prosperous suburb, woods, and fields, but little water surrounding it. From Calatfimi to the fortress of the Idol is twelve miles, we have already mentioned this place. From the fortress of the Idol to the Estate of the Commander [*Raḥl Al-Qā'id*] is ten miles.[68] Accordingly, from the Estate of the Commander to the Idols, which is on the coast is ten miles. The Delia [*Ṭūṭ*] river rises from *Jabal Hajar al-Ṣanam* and passes the Idol to the west then flows into the sea near Mazara.[69]

We'll return now and say that from Mazara to the Palace of Ibn Mankud [*Qaṣr Ibn Mankūd*] in a north-easterly direction is fifteen miles.[70] From the Palace of Ibn Mankud to Belice is four miles to the northeast.[71] From Belice to Santa Margherita Belice [*Manzil Sindī*] to the north-east is fifteen miles.[72] From Santa Margherita Belice to the Palace of Ibn Mankud is six miles. From Santa Margherita Belice to the Estate of

---

**67** The location of *Mīrjā* is uncertain.

**68** The location of *Raḥl Al-Qā'id* site is uncertain.

**69** *Jabal Hajar al-Ṣanam* means "The Mountain of the Rock of the Idol" and probably corresponds to the mountain of Salemi. Santagati, *La Sicilia di Al-Idrisi*, 110n168.

**70** The location of this site is uncertain though Santagati suggests Castelvetrano. Santagati, *La Sicilia di Al-Idrisi*, 111n170.

**71** Maurici places this at Castello della Pietra near the mouth of the Belice River. Maurici, *Castelli medievali in Sicilia*, 252.

**72** *Manzil Sindī* means "The Estate of Sindi." Previously named Misilindino, in the seventeenth century it became known as Santa Margherita Belice. Maurici, *Castelli medievali in Sicilia*, 323.

the Widower [*Raḥl al-Armal*] is nine miles north-west.[73] From Santa Margherita Belice to Calatamauro [*Qalʿat Mūrū*] is nine miles, and from here to Battalari is six miles to the east.[74] The Palace of Ibn Mankud is a vast estate which extends over a wide area in every direction. It is surrounded by gardens and farmland and has a rock that protects it.

Likewise, Belice is a strong castle, towering and well-protected. Mountains surround Belice on all sides and its rock protects it. There are a small number of trees and farms surrounding it, and the Belice [*al-Qārib*] River is nearby.[75] This river begins in the mountains that surround the fortress of Corleone to the north, then it flows east of Corleone, then turns westward arriving on the western side of Santa Margherita Belice . The river then flows between the mountains in a southerly direction and to the east of Belice, then continues southward, coming out at the sea near the Idols. The length of this river from its beginning to its end at the sea is fifty miles and from this river to the *Salmūn* River is five miles, and the *Salmūn* is a river that comes from a small mountain.[76] From the *Salmūn* River to Sciacca is twelve miles, and from Sciacca to Platani is seventeen miles.

Platani is a noble place, well-fortified. There is farmland, crops and many bounties; there are numerous orchards and trees. It has both settled and itinerant people and the Platani River flows along its eastern side. From Platani to *Ghardhūṭa* to the east is ___.[77] This is a substantial settlement with a big population. There are many orchards, trees and well-populated farmland. From *Ghardhūṭa* it is nine miles north to Sutera, which is surrounded by mountains on all sides. Sutera is prosperous and is a destination for incoming and outgoing travellers. It is eighteen miles north from the estate of Sutera to the castle of Cammarata which we have mentioned previously.

Likewise, from Agrigento to *al-Minshār* to the north-east is eighteen miles.[78] This is a castle on the top of a precipitous mountain. It has a large population, there is much farmland, and it is very fertile. From here to *al-Qiṭāʿ* to the south is ten miles.[79]

*Al-Qiṭāʿ* is an elevated settlement, situated on a mountain. There are crops, plenty of cultivation, it is abundantly fertile, a place where all benefits and advantages are combined. From *Al-Qiṭāʿ* to Agrigento is twelve miles westward and from *Al-Qiṭāʿ* to Platani is twenty miles to the north. From Agrigento to Naro is twelve miles to the east. Naro is a delightful estate, there are busy markets with active industry and there is a

**73** This is literally "The Estate of the Widow." Santagati places this in the area of Salaparuta. *La Sicilia di Al-Idrisi*, 113n174.

**74** *Qalʿat Mūrū* means "The Fortress of the Moor." The only other mention of this place is within a privilege granted by the Church of Monreale in 1182. Maurici, *Castelli medievali in Sicilia*, 264.

**75** *Al-Qārib* means "The Boat."

**76** Santagati posits that the *Salmūn* (literally "Salmon") River is the modern Garbo. Santagati, *La Sicilia di Al-Idrisi*, 114.

**77** Distance missing from all manuscripts. The location of *Ghardhūṭa* is uncertain.

**78** This is literally "The Saw [Tool]." Amari locates this at Canicattì, with some uncertainty. Amari, *Biblioteca Arabo-Sicula*, 1:96.

**79** This is literally "The Strip." The location of this site is uncertain.

market on a recognized market day. There are continuous tracts of farmland, and a great amount of cultivation. From Naro to *Al-Qiṭāʿ* in the north is ten miles and likewise from Naro to *Al-Sābūqa* in the east is twelve miles, it is the same direction to *Al-Qiṭāʿ*.[80] Likewise from *Al-Minshār* to *Al-Sābūqa* is eleven miles in a south-easterly direction.

*Al-Sābūqa* is an elevated castle and it is well-populated. There is much agriculture and different types of crops, its blessings are infinite and cultivation widespread. From *Al-Sābūqa* to Caltanissetta [*Qalʿat Al-Nisāʾ*] is twelve miles along the Agrigento Road.[81] From Naro to Caltanissetta to the north-east is twenty-one miles. Caltanissetta is a well-built fortress overlooking a continuous series of cultivated fields. It has excellent facilities, crops, trees and fruits and close by in an eastward direction flows the Salso River.

From Caltanissetta to Enna is eighteen miles. This is a city at the top of a mountain and it is one of the most impregnable castles. Its territory is vast and its construction imposing; there are charming, well-ordered markets, and dwellings of excellent construction. There are different crafts and merchandise, artisans and shops. It has vast areas of agricultural land, its farms are fertile, and its crops of good quality. The air is cool and the facilities of the town replenish the travellers who come and go. In summary, this is one of the most secure of Allah's regions, both for its situation and the solid quality of its construction. Along with its impregnability, upon its mountain there is farmland and running water, so there is no need of flat ground.[82] It has a towering castle, in an imposing position and it is not possible to overcome or defeat it.

Eighteen miles north from Enna is *Maḥkān*.[83] From *Maḥkān* to *Qaṣr* ___[84] is twenty-five miles south-east and from *Maḥkān* to Sutera to the west is twenty-five miles. Likewise from Sutera to Agrigento is thirty-six miles, this is a long day's journey and along the way you pass *Ghardhūṭa*, which we have already mentioned, then to *Al-Minshār*, then *Al-Qiṭāʿ* and then Agrigento. From Sutera to Castronovo to the north is twenty-four miles and we have mentioned these fortresses and strongholds previously in this book.

From Agrigento to *Qarqūdhī* to the east is one hundred and eighteen miles and from *Qarqūdhī* to Naro is twenty-four miles.[85] From Naro to Agrigento is twelve miles and likewise from Naro to Caltanissetta is twenty-one miles and from Caltanissetta to *Qarqūdhī* southward is fifteen miles. *Qarqūdhī* is a lovely region atop one of the most inaccessible mountains. It has fertile land and excellent, fecund cultivation.

---

**80** Maurici identifies *Al-Sābūqa* as Sabuci, but the exact location of the castle is uncertain. Maurici, *Castelli medievali in Sicilia*, 360.

**81** This is literally "The Fortress of the Woman."

**82** Meaning the town has all it needs despite the lack of flat land.

**83** Location uncertain. Bresc and Nef identify this as Micciché in the commune of Santa Caterina Villarmosa. Al-Idrisi, *La Première géographie de l'Occident*, 328.

**84** Name missing from all manuscripts.

**85** The location of *Qarqūdhī* is uncertain. Amari identifies this as Darfudi with some uncertainty, Amari, *Biblioteca Arabo-Sicula*, 99n2.

Within the lands of *Qarqūdhī* and lying on its eastern side is the Salso River. The source and origin of the Salso is at *Shaʿrā' Nizār* which is above *Jifla* and between *Shaʿrā' Nizār* and *Jifla* it is one and a half miles.[86] The river descends south opposite *Jifla* and between *Jifla* and the river is one mile. The river then passes Calathamet, then from there flows past an estate called *Ḥarāqa*, with this estate on the river's right, and between the river and *Ḥarāqa* is a stone's throw.[87] Between *Ḥarāqa* and Calathamet is six miles and up to this point the river's water is sweet. The river then flows until it reaches the territory of *Maḥkān*, with *Maḥkān* on the river's right, but before this the river flows through a swamp and its water becomes salty. The river then continues along the western portion of the territory of Enna and passes the eastern part of Caltanissetta at a distance of five miles, then travels two miles east of Pietraperzia [*Al-Ḥajjar Al-Mathqūb*].[88] The river then flows to the east of *Qarqūdhī*, which we mentioned previously, and between *Qarqūdhī* and the river is about nine miles. Then the river turns sharply west until it comes close to Licata and passes to the south of it, then flows into the sea at a very short distance from that city.

From *Qarqūdhī* to Butera to the south is twelve miles through the mountains, and if you don't pass through the mountains it is twenty-four miles, we mentioned Butera previously. From Butera to Licata is nineteen miles, we mentioned Licata previously in the description of the coastal cities. Between Butera and *Shalyāṭa* is twelve miles to the north-east.[89] *Shalyāṭa* is an estate on a plain with gushing rivers, fertile farmland, plentiful bounties and numerous crops. The *Al-ʿAsal* river flows in the western part of *Shalyāṭa*'s territory and between *Shalyāṭa* and Piazza Armerina [*Iblāṭsa*] in a northerly direction is ten miles.[90] The aforementioned *Al-ʿAsal* River descends from Piazza Armerina.

Piazza Armerina is a secure castle with expansive territory, its farmland is bounteous. There is a famous market there with many crops, trees and fruits. From Piazza Armerina to *Qarqūdhī* to the west is around fifteen miles and from Piazza Armerina to Pietraperzia is about the same distance. Pietraperzia is a strong fortress, a robust stronghold, its territory is extensive and well-populated and there is much water. From Pietraperzia to Enna is about twelve miles and likewise from Pietraperzia to *Shalyāṭa* is twenty-five miles. Caltanissetta is north-west of Pietraperzia at a distance of seven miles.

Between *Shalyāṭa* and *Ḥiṣn Al-Janūb* which is called Caltagirone [*Qalʿat Al-Khinzāriyya*] is ten miles, this is a secure fortress on the side of an inaccessible mountain.[91]

---

**86** This is literally "The Woods of Nizar." The location is uncertain.

**87** This is literally "Burning/fiery." The location is uncertain.

**88** This is literally "The Pierced Stone."

**89** Location uncertain. Bresc and Nef identify *Shalyāṭa* as Garsiliato. Al-Idrisi, *Première géographie de l'Occident*, 329.

**90** This is literally "The Honey River." Santagati suggests this is the Dissueri River. Santagati, *La Sicilia di Al-Idrisi*, 130n209.

**91** This is literally "The Fort of the South." Amari misidentifies the ب with a ن causing the mistranslation of "The Fort of the Spirits." *Qalʿat Al-Khinzāriyya* is literally "The Fort of the Boars."

Its excellent territory is vast, its cultivation widespread and there is much honey there. Between Caltagirone and Ragusa is twenty-five miles. Ragusa is a pleasant estate, well-constructed, elevated and very secure.[92] Ragusa is located on a river which is its namesake and between Ragusa and the sea is twelve miles and then to Scicli is twelve miles eastward. It is eight miles between Scicli and Modica and from Ragusa to Modica is five miles northward.

Modica is located between impenetrable mountains; it is a bounteous place with various benefits and plentiful crops. Between Modica and Buscemi [*Qalʿat Abī Shāma*] to the north is sixteen miles, between Buscemi and Ragusa is fifteen miles to the south and between Buscemi and Lentini is twenty-four miles.[93]

Buscemi is a stronghold, it is dependable as a place of refuge and is surrounded by woods; both the Eloro[94] [*Al-Arwā*] and Anapo [*Bantārgha*] Rivers descend from its mountains. The Eloro River ends at Syracuse's port and the Anapo flows into the sea at the southern corner of the island. Between Lentini and Vizini is twenty-five miles south-west and likewise from Ragusa to Vizzini is twenty miles, between *Shalyāṭa* and Vizzini is twenty-five miles.

Vizzini is at the foot of a mountain and it has farmland of good quality. Two rivers descend from its mountains—they come out separately and then unite at some distance from Vizzini progressing through the mountains and the forest to the sea; this river is called the Dirillo [*Ikrīlū*]. Between Vizzini and Buscemi is fifteen miles and between Buscemi and Noto is thirty miles. Between Noto and the sea in the direction of Malta is twenty miles. Between Noto and Pantalica is nineteen miles.[95] Pantalica is surrounded by Syracuse's mountains and the river that is its namesake starts in the area of Buscemi which we mentioned previously. Between Pantalica and Syracuse to the east is nineteen miles and between Pantalica and Lentini is twelve miles west. Between Pantalica and Mineo to the south-west is twenty-four miles.

Mineo is a pleasant fortress situated in the mountains of Vizzini and it is the source of many springs. There is plenty of farmland, fruits and dairy products, and the land has good soil. Between Mineo and Vizzini is fourteen miles southward and from Mineo to Caltagirone is ten miles to the west and from Mineo to *Qalʿat Al-Fār*[96] is three miles to the north and between Mineo and *Manzil Maljāʿ Khalīl* is nine miles.[97] *Manzil Maljāʿ*

---

**92** Ragusa is described twice—once in the coastal settlements and once in the interior.

**93** This is literally "The Fortress of Abi Shāma."

**94** The Eloro is also known also as the Abisso or Tellaro river. Vincenzo Garofalo, "Nuova ipotesi sull'etimologia di 'Tellàro', il fiume che sbocca a sud di Siracusa," *Incontri: La Sicilia e l'altrove* 4 (July–September 2013): 9–11. The Anapo River was previously known as Fiume di Pantalica. Santagati, *La Sicilia di Al-Idrisi*, 135n228.

**95** The Late Bronze-Age necropolis of Pantalica. Robert Leighton, "Pantalica (Sicily) from the Late Bronze Age to the Middle Ages: A New Survey and Interpretation of the Rock-Cut Monuments," *American Journal of Archaeology* 115 (2011), 447–64.

**96** This is literally "The Fort of the Mouse." Maurici identifies this as ruins located at Monte Catalfaro. Maurici, *Castelli medievali in Sicilia*, 138–39.

**97** This is literally "The Estate of the Shelter/Refuge of Khalil." Al-Idrisi also refers to this as *Manzil*

*Khalīl* has extensive cultivation and agriculture. There is a mountain to the south where a river called the *Būkrīṭ* begins.[98] Between *Manzil Abī Khalīl* and Caltagirone is nine miles to the south and between *Manzil Khalīl* and Enna is twenty-four miles. From Mineo to Buccheri is eighteen miles east along the mountain road.

Buccheri is an estate situated on a plain. It is a prosperous and pretty place with a good yield of crops and an abundance of fruit. To the west, Buccheri connects to the pine forest known as *Al-Banīṭ*.[99] From Buccheri to Lentini to the north is twenty miles and from Buccheri to Buscemi to the south is seven miles and the two territories are contiguous.

From Enna to Piazza Armerina to the south is twenty miles.[100] Piazza Armerina is a castle between Caltagirone and Pietraperzia and between Piazza Armerina and Pietraperzia is fourteen miles. And like this, between Piazza Armerina and *Shalyāṭa* to the south is twelve miles and between *Manzil Abī Khalīl* and Paternò is twenty miles. Between Buscemi and Palazzolo Acreide is two miles and between Palazzolo Acreide and *Qīrī* is twenty-two miles.[101] From Piazza Armerina to Aidone is nine miles to the north.

The Gornalunga [*Ranbalu*] River springs from Aidone and then flows eastward where it meets the *Būkrīṭ* River that we mentioned previously. The two rivers then flow into the Dittaino [*Al-Ṭīn*] River eight miles from where the two rivers first met.[102] Moving towards the sea, the rivers combine with the Simeto River, becoming one river and emptying into the sea.

Between Aidone and Enna is fifteen miles north-west and likewise from Aidone to *Manzil Abī Khalīl* is around ten miles, and from Enna north to Tavi is ten miles.[103] Tavi is a splendid and elevated castle; it has farms and water. The Dittaino River rises there and flows eastward until it meets the Simeto River near the sea.

From Tavi to Castel di Iudica is twelve miles to the east and from Aidone to Castel di Iudica is also twelve miles, northward. Castel di Iuidica is a large estate with a great number of people, its farmland is extensive and there are many useful crops. From Castel di Iudica to *Manzil Abī Khalīl* to the south is thirteen miles and from Tavi north to Agira [*Shant Fīlib*] is eleven miles and from Agira to Centuripe is fifteen miles.[104]

---

*Abi Khalīl* (The Estate of Khalil's father). Santagati locates this at the ruins of Mongialino. Santagati, *La Sicilia di Al-Idrisi*, 139n244.

**98** Amari identifies this as the Margarito Rver; however, Santagati identifies this as the Ferro, an affluent of the Gornalunga River. Santagati, *La Sicilia di Al-Idrisi*, 140n245.

**99** The location of *Al-Banīṭ* is unknown.

**100** Al-Idrisi describes the settlement of Piazza Armerina twice.

**101** Santagati locates this at Contrada Sciri, near Mazzarone. Santagati, *La Sicilia di Al-Idrisi*, 143n258.

**102** This is literally "The River of Clay."

**103** The ruins of Castello di Tavi are located on Pizzo Castellaccio, near Leonforte. Maurici, *Castelli medievali in Sicilia*, 375.

**104** This settlement was known as San Filippo d'Argiriò until 1861. Girolamo Caracausi, *Dizionario onomastico della Sicilia*, vol. 2 (Palermo: L'Epos, Centro di studi filologici e linguistici siciliani, 1994), 19.

Centuripe is a lovely place with many advantages and its crops are extensive, it is well-populated in every direction and its lands stretch far and wide. Centuripe is east of Agira. Agira is one of the loveliest and most noble places; it has amongst the largest number of crops and abundance of any place. Between Centuripe and Adrano is thirteen miles northward. Above Adrano, the Troina, Cerami and Capizzi Rivers and other smaller rivers meet.

Adrano is a lovely estate much like a small city, located on a high rocky point at the southern foot of Etna. There is a market, a bathhouse and a good fortress with plenty of water. . From Adrano along the foot of the mountain to Paternò is six miles.

Paternò is an impenetrable stronghold and castle; its benefits are numerous, there is plenty of farmland, fruits, grapevines and gardens. It is a lovely fort overlooking its territories and between Paternò and Sant'Anastasia is seven miles south-east. Between Motta Sant'Anastasia and the sea is twelve miles and between Motta Sant'Anastasia and Lentini to the south is nineteen miles. Between Motta Sant'Anastasia and the Simeto River is two and a half miles.

The Simeto River comes from four sources. The first of these is the Cerami River whose source is in the mountains of Capizzi, the second source also comes from these mountains and its gardens. The Cerami River flows between the two mountains for two and a half miles, then combines with the second river before proceeding on to Cerami and between the confluence of the two rivers and Cerami is about six miles. The river passes below Cerami, where the mills are situated and Cerami is located to the east of the mills at a distance of one mile. Between the confluence of the two rivers and *Ḥajar Sārlū* is eight miles and located here is the Nicosia River and between the Nicosia and the Cerami Rivers is one big mile.[105] From there the now united river descends between Agira and Gagliano. Gagliano is to the east at a distance of one and a half miles and Agira on its western side at a distance of half a mile. The river then descends to *Intir Nistīrī* located between Aderno and Centuripe, with Aderno on the river's eastern side at a distance of one mile and Centuripe on the western side of the river at a distance of one and a half miles.[106] This river then meets the Simeto River at the aforementioned place combining with the river that descends from Troina together with the Elia and Ambola Rivers.

From Troina to the confluence of the aforementioned rivers is eight miles and from the Elia to this point is four miles, from the Ambola to this point is five miles. The rivers then combine and become one, descending to *al-Jurṭa* and continuing leaving Paternò and Motta Sant' Anastasia on the east side and between Paternò river is half a mile.[107] Between Motta Sant'Anastasia and the Simeto river for two miles. Finally, all

**105** This is literally "Serlo's Rock", referring to Serlo II of Hauteville, nephew of Roger I, who in 1072 died fighting Muslim forces. Santagati, *La Sicilia di Al-Idrisi*, 149n285. The location is uncertain.

**106** The location of *Intir Nistīrī* is uncertain. Bresc and Nef locate it at Monte Vaccara near Regalbuto. Al-Idrisi, *Première géographie de l'Occident*, 334n1.

**107** Amari suggests *al-jurṭa* could be "ship/boat." Amari, *Biblioteca Arabo-Sicula*, 1:111n2. The location is unknown.

the rivers come together—the Simeto, the Dittaino, *Ranblū* and *Karīṭ* near the sea then flow into it.[108]

We will return now and say that from Viccari to Petterana is nine miles and from Petterana to Sclafani Bagni is five miles. From Sclafani Bagni to Caltavuturo [*Qalʿat Abī Thūr*] to the east is six miles.[109] Caltavuturo is an impenetrable castle; it is well-populated, has good farmland and abundant crops.

From Caltavuturo to Polizzi Generosa in the south is five miles. Polizzi Generosa is an elevated castle located in one of the most beautiful locations and it has fertile farmland. From Polizzi Generosa to Petralia Sottana to the east is six miles. Petralia Sottana is a noble castle, a secure stronghold, it has expansive farmland and its products and blessings are numerous. There is a market and a fort; the market is like markets in big cities. From Petralia Sottana to *Maqāra* is eight miles, this is a well-populated castle with many dwellings, plentiful farmland and benefits.[110] From *Maqāra* to the fortress of Sperlinga is ten miles south, Sperlinga is a large estate where everything is good whether it be land or agriculture and its cultivated fields are extensive. From Sperlinga to Cammarata is twenty-three miles, we have previously mentioned Cammarata.[111] From Sperlinga to Nicosia to the east is twelve miles.

Nicosia is one of the greatest of all castles, it has a well-populated suburb and extensive cultivated land and from Nicosia to Troina is twelve miles north-east. Troina is an elevated castle much like a city; it is a well-settled place, surrounded by farms and cultivation. From Troina westward to Cerami is eight miles. Cerami is an estate at an elevated position on a rock, it is well-populated, its crops are fertile and there is plenty of freshwater. From Cerami to Capizzi is nine miles northward.

Capizzi is a castle of robust structure and prosperity, it is well-fortified and has many vineyards and a profusion of excellent attributes. From Capizzi to Geraci Siculo is fifteen miles westward. Geraci Siculo has an abundance of fruits, it is prosperous, its farms and village are large and its cultivation widespread. Geraci Siculo is situated between towering mountains and a succession of peaks and between Geraci Siculo and Petralia Sottana is about ten miles and between Geraci Siculo and *Ruqqa Bāsīlī* northward is nine miles.[112] *Ruqqa Bāsīlī* is a lovely place where it is possible to make a good living, it has many advantages and excellent cultivation.

From *Ruqqa Bāsīlī* to Isnello [*al-Ḥimār*] which is an estate at the top of a mountain, it is ten miles westward.[113] From Geraci Siculo to Isnello is thirteen miles and from Isnello to Polizzi is six miles south-west. From Isnello to Collesano [*Qalʿat al-Ṣirāṭ*] is nine miles westward, this is a fortress located on an inaccessible hill, there is plenty of

**108** Previously al-Idrisi writes *Karīṭ* as *Būkrīṭ*.

**109** This is literally "The Fortress of Abi Thūr."

**110** The location of *Maqāra* is uncertain.

**111** This distance is way off, it is around 100 modern kilometres or 190 Arabic miles.

**112** This is literally "Basil's Rock" or "The Rock of the King." Santagati places this at Monte Molucco. Santagati, *La Sicilia di Al-Idrisi*, 158n32.

**113** This is literally "The Donkey."

water and farmland and there is an imposing mountain overlooking it.[114] In the past there was an impenetrable castle on this mountain which was the epitome of safety and protection but which now is a grazing place for sheep and cows. Then the great King Roger had the fortress demolished and erected the fort that is located there now. From Collesano to Cefalù on the coast is eight miles and between the two is Gratteri, this is a small fortress with many advantages. From Collesano to coastal Termini is fifteen miles westward and one mile to the north. From *Ruqqa Bāsīlī* to Tusa which we mentioned previously is ten miles to the north.

We will return now and say that from Troina which we have already mentioned to Maniace is twenty miles, the latter is also known as *Ghīrān al-Daqīq*.[115] This is a prosperous village on a plain, there is a market and merchants, great fertility and abundant blessings. Maniace is located at the northern corner of the mountain called Etna and between the two it is nearly five miles. Maniace is located on a river with many mills whose source is about three miles distant. It is twenty miles from Maniace to Aderno via the road running beside the Simeto valley and we mentioned Aderno previously. From Maniace eastward to Randazzo is ten miles. Randazzo is at the bottom of the aforementioned mountain and is a village resembling a small city, its market bustles with traders and artisans and there is a great amount of wood available which is exported far and wide. From Randazzo to Castiglione di Sicilia is twenty miles and between the two there is a castle resembling an estate called Mojo Alcantara [*al-Mudd*].

Castiglione is elevated, impregnable and well-populated, its markets are busy with buying and selling. From Castiglione to the village of Mascali which is located on the side of the mountains that face the sea ___.[116] Mascali is a well-populated village on a hill and water flows through the middle of it. From Mascali to Taormina on the coast is six miles and between the two is *al-Bārid* river which comes out from the tall mountains to the west of Maniace, flowing eastward continuously until it reaches the sea; the full course of this river is eighty miles.[117]

And from Randazzo to Montalbano Elicona is twenty miles. This is a fortress located between imposing mountains whose ascent and descent are both difficult, it has livestock, honey and many bounties. From Montalbano Elicona to *Manjaba*___.[118] From *Manjaba* to Galati Mamertino to the west is ten miles. This is an impregnable fortress located between imposing mountains, it is prosperous and well-populated and there is farmland, livestock and a great amount of flax is cultivated using irrigation.

From Galati Mamertino to the church of San Marco is seven miles north-west.[119] From San Marco to San Fratello is five miles and from here to Caronia is fourteen

---

**114** This is literally "The Fortress of the Path."

**115** This is literally "The Caves of Flour."

**116** Distance missing from all manuscripts.

**117** *Al-Bārid* is literally "The Cold River."

**118** The location of *Manjaba* is uncertain. Santagati believes it could be Floresta. Santagati, *La Sicilia di Al-Idrisi*, 165n347. Distance missing from all manuscripts.

**119** Al-Idrisi describes *Shant Mārkū* as a *qalʿa* (fortress) in the coastal itinerary and as a *kanīsa* (church) in the interior.

miles.[120] Caronia is a castle on top of a mountain, overlooking the sea where tuna is fished, there are vineyards and cultivated fields. From Caronia to *Qalʿat Al-Qawārib* is nine miles and between *Qalʿat al-Qawārib* and the sea is two miles. From *Qalʿat al-Qawārib* to the port of Tusa is seven miles and from Tusa to Cefalù is twelve miles.

We will return now and say that from Messina to the fortress of Rometta is nine miles. From Rometta to Monteforte San Giorgio is four miles southward. From Monteforte San Giorgio to Milazzo is fifteen miles northward and from Monteforte San Giorgio to *Mīqash* southward is fifteen miles, and *Mīqash* is located between Messina and Taormina on a difficult road.[121]

From Locadi to Tripi [*Barbalis*] is fifteen miles north-west and Monteforte San Giorgio to Tripi is twenty miles westward. Tripi is a well-built and spacious fortress and its people make a good living. From Tripi to Mojo Alcantara is five miles southward and from Tripi to Montalbano Elicona is twelve miles and like this from Montalbano Elicona to Mojo Alcantara is ten miles.

Here Sicily has been described and there is no island in the sea that has a greater number of towns or populated regions. All that remains is to list the island's ports, anchorage by anchorage, and distances between them, both in miles and days' journey, by the will of Allah almighty:

And we say from the city called Palermo travelling along the extent of the bay to Vergine Maria is five miles. From Vergine Maria to Mondello is five miles and from Mondello to Capo di Gallo is two miles. From Capo di Gallo to Isola delle Femmine is four miles. From there to Carini is six miles. From Carini to the promontory which is below Terrasini is three miles. From there to the creek of Terrasini is three miles and from there to the promontory which is between it and Partinico[122] is three miles. From there to the beach that is below Partinico is one and a half miles. From there to the Calatubo River is five miles and from there to the Castellammare del Golfo River is a quarter mile. From there to the mountain of Santo Vito is twelve miles. From Santo Vito to Trapani is twenty-five miles and from Trapani to Marsala is twenty-five miles. From Marsala to Capo Feto which is between Marsala and Mazara is twelve miles, from Mazara to Capo Granitola is six miles and from there to Tre Fontane is six miles. From Tre Fontane to the Idols is four miles and from there to Porto Palo is six miles. From Porto Palo to the Carabo/Garbo River is six miles and from there to Capo San Marco is six miles.

From Capo San Marco to Sciacca is six miles and from there to the *Albū*[123] River is eight miles. From the *Albū* River to the mouth of the Platani River is nine miles. From the mouth of the Platani River to the Anchorage of ʿAbād[124] is six miles. From

**120** Al-Idrisi describes Caronia twice.

**121** The location of *Mīqash* is uncertain. Santagati suggests Monte Dinnamare. Santagati, *La Sicilia di Al-Idrisi*, 168n361.

**122** The Capo di Rama.

**123** This was previously the Allaba or Allava River, now known as the Bellapietra River near Sciacca. Luciano Rizzuto, *Camico. Topografia di una fortezza* (Agrigento: Estero Salvatore, 2004), 41.

**124** Amari places this at Punta di Caribici, but according to Santagati this is more likely Siculiana Marina. Santagati, *La Sicilia di Al-Idrisi*, 173n391.

the Anchorage of ʿAbād to Monte Rossello is nine miles, from there to Agrigento is nine miles. From Agrigento to the *al-Zakūjī*[125] River is three miles. From there to Pietra Padella is nine miles and from there to *Baswārīyya*[126] is eighteen miles. From *Baswārīyya* to the Saltworks[127] is three miles from there to Licata is three miles.

From Licata to the Salso River is one mile and from the Salso to Falconara is eight miles. From there to Manfria is eight miles and from there to the Masts[128] River is twelve miles. From the Masts River to the Greek River[129] is twelve miles. From the Greek River to the Island of the Dove is twelve miles and from there to the *Karnī*[130] River is seven miles. From the *Karnī* River to the Erminio River is twelve miles. From the Erminio River to the cliff of the Child[131] is four miles and from there to the Port of Scicli[132] is four miles. From the Port of Scicli to the Marina di Ragusa is two miles and from there to the Port of the Dromon[133] is four miles. From the Port of the Dromon to the Port of the Tree[134] is one mile. From the Port of the Tree to Isola dei Porri is three miles. From Isola dei Porri to La Marza is three miles and from La Marza to Isola dei Correnti is eight miles. From La Marza to Portopalo [*Karam al-Ranbūḥ*][135] is three miles and from Portopalo to the Cape of Bashnu[136] is three miles. From the Cape of Bashnu to the Entrance of the Bowls[137] is six miles. From the Entrance of the Bowls to Marzamemi is six miles and from Marzamemi to Vendicari is six miles and from here to *al-Qāṭa*[138] is six miles.

From *al-Qāṭa* to the Cassibile River is twelve miles and from the Cassibile River to the Port of the Expert[139] is six miles and from there to Ognina is six miles. From Ognina

**125** Santagati states this could be either the Drago, San Biagio or Akragas Rivers. Santagati, *La Sicilia di Al-Idrisi*, 173n394.

**126** Amari places this at Punta San Nicola but Santagati proposes Torre di Gaffe. Santagati, *La Sicilia di Al-Idrisi*, 174n396.

**127** Amari places this at Rocca Muddafedda, Santagati states this could be Mollarella or Mollachella. Santagati, *La Sicilia di Al-Idrisi*, 174n397.

**128** Santagati suggests the Gela River, *La Sicilia di Al-Idrisi*, 174n401.

**129** Santagati states this could be either the Dirillo or Acate Rivers. Santagati, *La Sicilia di Al-Idrisi*, 174n402.

**130** Santagati proposes that this is the river of Ippari. Santagati, *La Sicilia di Al-Idrisi*, 175n404.

**131** Santagati proposes Punta Bruga, just west of Cava d'Aliga. Santagati, *La Sicilia di Al-Idrisi*, 175n406.

**132** Santagati proposes Sampieri. Santagati, *La Sicilia di Al-Idrisi*, 175n407.

**133** Location uncertain. The dromon was the most important byzantine warship from the fifth to twelfth centuries. John Pryor and Elizabeth Jeffreys, *The Age of the Dromon: The Byzantine Navy ca. 500–1204* (Leiden: Brill, 2006).

**134** Santagati proposes this to be the stretch of beach delimited by the Salaia and Favara Rivers. Santagati, *La Sicilia di Al-Idrisi*, 176, 410.

**135** This is literally "The Vineyard of *Ranbūḥ*."

**136** Santagati proposes the point south of Portopalo Santagati, *La Sicilia di Al-Idrisi*, 176n415.

**137** Location uncertain.

**138** The location of *Al-Qāṭa* is uncertain.

**139** Location uncertain.

to Capo Murro di Porco is eight miles and from there to Syracuse is six miles. Then to Capo Santa Panagia is six miles and then to Penisola Magnisi is four miles, then to Porto Xifonio is four miles. From Porto Xifonio to Santa Croce is six miles and from there to the Brucoli River is six miles, then to Agnone is six miles and from there to the San Leonardo River is three miles, then to Catania is six miles. Then to Ognina is three miles, then to the Cyclopean Isles is three miles and then to Capo Molini River is three miles. From there to Santa Tecla is six miles and then to the Spring of the Reeds[140] is three miles. Then to the Capo di Mascali is three miles and then to the Alcantara River is nine miles, and then to Capo Schisò three miles. Then to *al-Anbāṣī*[141] is five miles and then to the Step[142] is ten miles and from here to *Shant Bālmī*[143] is five miles and from *Shant Bālmī* to Marina di Fiumdinisi is six miles and from there to the Middle Step[144] is six miles. From there to the Spring of the Sultan[145] is two miles and from there to Scaletta Zanklia is two miles. From Scaletta Zanklia to Briga is three miles and from there to Santo Stefano di Briga is three miles.

From Santo Stefano di Briga to Tremestieri is seven miles and from there to Messina is six miles. From Messina to Capo Peloro is twelve miles and to the Saponara River is twelve miles. To Milazzo is twelve miles and to the cape of Milazzo is six miles. From the cape of Milazzo following the gulf is twenty-five miles to Oliveri and to the cape of Tindari is three miles. To Patti along the gulf is four miles and to Capo Calavà is two miles. To Brolo is four miles and to Capo di Orlando is three miles, then to the beach of Acquedolci is twenty-six miles. To Caronia is twelve miles and to Mistretta is six miles and then to Tusa is six miles. To Capo Raisgerbi is four miles and along the gulf to Cefalù is eight miles. To *Ḥijr ʿAmmār*[146] is two miles and to Capo Plaia is four miles, to Roccella is six miles and to the Masts River is three miles and to the Torto River is three miles to Termini is six miles and to Trabia it is three miles. To Solanto is three miles and to the village of Aspra is six miles. To the Misilmeri river along the gulf is two miles and finally to Palermo is six miles.

We have described the island of Sicily and now to describe its shape: Sicily is triangular in shape and at the eastern point is the city of Messina and from there to Capo Passero is two hundred miles and from Capo Passero to Trapani is four hundred miles and this is the southern side of the island and the third side of the island, from Trapani to *al-Ḥarāsh*[147] and to Capo Peloro is two hundred and fifty miles and here we finish our discussion.

We have completed Part Two of Clime Four, thanks be to Allah, and now we go on to Part Three Allah willing.

---

**140** Location uncertain.

**141** The location of *Al-Anbāṣī* is uncertain.

**142** Location uncertain.

**143** The location of *Shant Bālmī* is uncertain.

**144** Location uncertain.

**145** Location uncertain.

**146** This is literally "The Stone of ʿAmār." Location uncertain.

**147** The location of *al-Ḥarāsh* is uncertain.

Figure 4. Map of Sicily and the Mediterranean (Oxford, Bodleian Library, MS Pococke 375, fols. 187b–188a).


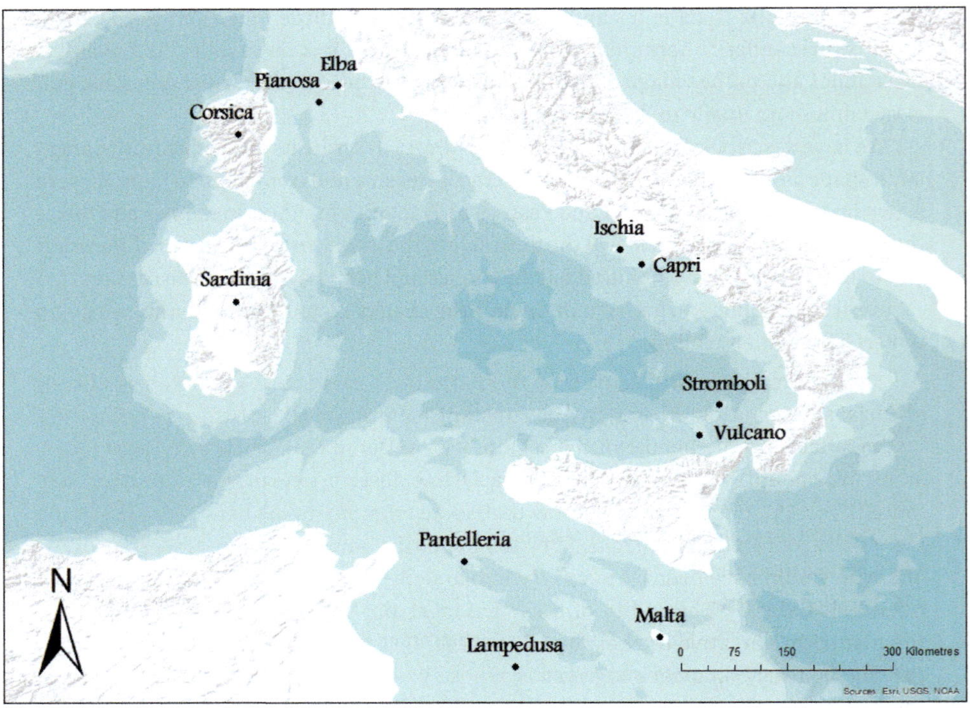


Map 2. Central Mediterranean Islands. Map by authors.

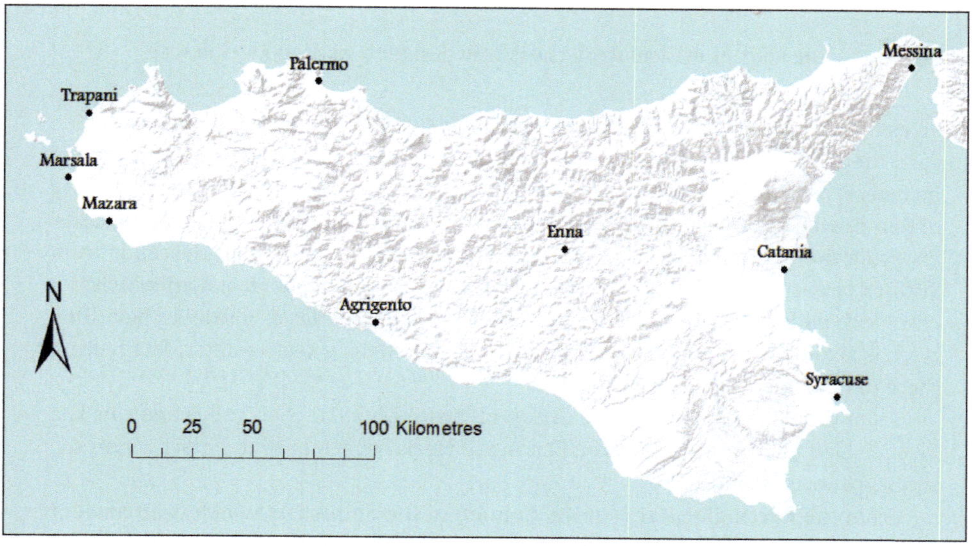


Map 3. Cities in Sicily. Map by authors.

## CLIME FOUR, PART THREE

THIS THIRD SECTION of the Fourth Clime includes a portion of the Mediterranean Sea from the islands of Corfu, Lefkada, Fano, Cephalonia and the island of Zante. And also in this section there are both coastal and inland settlements such as Sant' Eufamia, Tropea, *Al-Māṣa*,Tuccio, Gerace, Nardo, Gallipoli, Castro, Otranto, Brindisi, Lecce, Vlorë [*Lablūna*], Butrint, Himarë, Parga [*Fāskiyū*], Vonitsa, Drinpoli and Ianina.

It now falls to us to describe the places mentioned, town by town and region by region by Allah's will. And we say that this sea from its western side measures six long days of sailing, travelling from Reggio to Gabès directly.

Reggio is a city in Calabria on the bank of the straits to Sicily. Between Reggio and Messina on the island of Sicily is seven miles and the strait is the extent of the distance between the two cities. Reggio is a small and civilized city which has abundant fruit and legumes. There are busy markets, bathhouses and its walls are made of stone. The city wall is on the sea on the eastern side of the strait.

And from Reggio along the coast to Capo Pellaro is six miles.

Between Reggio and Bruzzano is a gulf called Bruzzano's Gulf and it is located at the head of the strait. From Capo Pellaro to Patarriti is three miles and from there to Pentedattilo moving inland is three miles. From Pattarriti to the Melito river [*Wadi al-ʿAsil*][1] is six miles and from there to the village of *Ṭābla* is six miles.[2]

From *Ṭābla* to cape Zeffirio is twelves miles and from there to *Baṭraqūqa* which is a running river, is three miles.[3] From *Baṭraqūqa* to Bruzzano is six miles. Bruzzano is

---

**1** This is literally "The Honey River."

**2** Bresc and Nef suggest that *Ṭābla* may be Capo San Giovanni. Al-Idrisi, *La Première géographie de l'Occident*, 341n2.

**3** Amari and Schiaparelli propose the Petracucca River with some reservations. Al-Idrisi, *L'Italia descritta nel "Libro del Re Ruggero"*, 72n2.

an estate on a mountain, there are cultivated fields and fertile territory and its people earn their living through sheep and cows. There are abundant crops and steady tax income. From Bruzzano to the Gerace River is twelve miles and on this river is the city of Gerace which is a large and pleasant city with cultivated fields, crops and vineyards.

From the Gerace River to the Alana River which is the river that descends from Stilo, is twenty-four miles and from the Alana to the Ancinale [*Baḥlāya*] which is a river with mills upon it, is twelve miles. From there to the Alessi, which is a navigable river, is sixteen miles, and from there to the Tacina River is twelve miles, finally from there to the Simeri River is twelve miles.

From the Simeri River to Isola di Capo Rizzuto [*Awsala*] is six miles and this is a lovely island and from there to the Earring of Maria[4] which is a large port where sea squill grows, is six miles.[5]

From the Earring of Maria to the Column of the Shipmasts which is an ancient settlement, is six miles.[6] From there to the city of Crotone which is a port and is of ancient construction, in a pleasant region and well-populated, is ten miles. And from Crotone to the port of the San Severina River is twelve miles and this is a port which is sheltered from the three winds, from San Severina to Punta dell'Alice is twenty-four miles.

And from Punta dell'Alice to the church which is at the cape of Fiumenicà [*Rās Abrāqna*] is twelve miles and from there to Rossano is twenty miles and from Rossano to the Crati [*al-Wādī al-Kabīr*][7] River is twelve miles and from there to Roseto is twelve miles and from there to Pietra di Roseto [*Ṣakhrat Sakan*][8] is twelve miles and this rock marked the border between the Franks and the Lombards.

From Pietra di Roseto to the Sinno River is six miles, and this is a navigable river with good anchorages. From the Sinno to the Bradano River is twenty-four miles and on this river and the mountains that surround it are plentiful pine trees and both pitch and tar are produced here and exported to many regions.

And from the Bradano River to the Agri River is six miles and from there to the Basento [*Al-Mawjan*] River is eighteen miles. From the Basento River to the Bradano [*Abrāghna*][9] River is three miles and from there to the Lato River is fifteen miles. From there to the Lenna River is three miles and from there to the Tara [*al-Muʿwajj*][10] River is six miles and from there to the Mills of Taranto[11] River is three miles and from there to Taranto is six miles.

---

**4** Location uncertain.

**5** *Drimia maritima* also known as sea onion or maritime squill.

**6** Location uncertain.

**7** This is literally "The Big River."

**8** This is literally "The Rock of Sakan"; also known as the Castrum Petrae Roseti.

**9** Al-Idrisi uses a different name for the same river, the Bradano.

**10** This is literally "The Curved."

**11** Location uncertain.

The city of Taranto is large and ancient with well-constructed buildings and homes. Many merchants and visitors visit the city, ships depart from it and companies of travellers visit it.[12] There are many shops and much wealth. There is a port on the open sea on the western side of the city and on the north-eastern side of the city there is a lake. The circumference of the lake from the bridge to the city gate is twelve miles. The bridge is situated between the open sea and the lake; the length of the bridge from the city gate to the mainland is three hundred cubits and its width is fifteen cubits. Within this bridge there is an opening which empties water from the open sea to the lake and vice versa two times during the day and two times at night. Three rivers empty into the lake and the depth of the lake is between ten, twenty-five and thirty fathoms. Taranto is surrounded by open sea and lake on all sides except its northern side.

From the city of Taranto twelve miles in an eastward direction is the Saturo and this is a river where boats anchor. From here to the *Aqlūya*[13] River is three miles and from there to the port of Colimena [*Qīṭūlqa*] where there is a spring is twelve miles. From Colimena to the large port of Cesarea [*Majlūd*] which is protected from winds from every direction and surrounded by wells and cisterns, is twelve miles. From there to the small port of Nardo, also known as *Nāṭrus*, is three miles. Between Nardo and the sea is four miles and from Nardo to Santa Maria al Bagno [*Ṭarāja*] which is a pleasant port with much good water is twelve miles. From Santa Maria al Bagno to the city of Gallipoli is six miles.

And the city of Gallipoli is large, old and prosperous. The city is located at a position on the land where the sea surrounds it as if it were an island. From Gallipoli to Torre S. Giovanni [*Anbāna*] which is a port located on a freshwater spring, is fifteen miles. From there to Leuca is fifteen miles, Leuca is on a promontory which overhangs the sea and there are two freshwater springs located at its port. From there to Castro is ten miles and this is a small city on the coast and from there to the city of *Rūthmart* heading inland is twelve miles.[14] From *Rūthmart* to Soleto is three miles and from there to the city of Lecce is twenty-six miles, from Lecce to the city of Otranto is twenty miles.

The city of Otranto is a city of ancient monuments and is very populous, and there are prosperous markets and lively commerce. The sea surrounds the city walls on three sides and the city is connected with the inland on its northern side. There is a river which descends from the north, then flows past the city gate and empties into the Adriatic Sea at Brindisi. There are forty miles between Brindisi and Otranto. The inhabitants of Brindisi are Lombards and previously the city was under the control of the lord of Constantinople.

---

**12** Al-Idrisi uses the word "*rifāq*", meaning "comrades/companions" but also "company of travellers." Amari and Bresc and Nef have translated this as "caravans."

**13** This river is unknown.

**14** Bresc and Nef place this at the Roman town of Rudiae Al-Idrisi, *La Première géographie de l'Occident*, 344, note 1 while Amari and Schiaparelli place it at Monteroni di Lecce. Al-Idrisi, *L'Italia descritta nel "Libro del Re Ruggero"*, 76n2.

The city of Otranto is situated on the western side of the strait between the Mediterranean Sea and the Adriatic. From Otranto to the city of Durrës is seventy miles.

Durrës is a large and prosperous city with many advantages. It has numerous markets, commerce and abundant luxury. The city is situated on the eastern side of the strait. From Durrës to the city of Vlorë which is on the Adriatic Sea is one hundred and twenty-five miles following the coastline of the gulf. From Vlorë to Himarë following the coastline is one hundred miles and the Greek name for the city is *Badnatū*.[15] And from there to Butrint is forty miles, Butrint is a small and civilized city with markets and prosperity.[16]

Himarë is a large, prosperous city and from Himarë along the coast to the city of Parga is sixty miles, Parga is a medium sized city and on its eastern side is a small gulf which resembles an arm with boats entering through a narrow opening.

From Parga to the city of Vonitsa is thirty miles, or twenty miles in a straight line. Vonitsa is a medium sized city which is prosperous, it has a city wall and a market where people buy and sell, and from there to Lepanto is one hundred and fifty miles. Out to sea off Himarë is an island called Othonoi and between the two there is a distance of thirty miles. Othonoi is a pleasant island which has a port and woods.

And west and slightly to the south of this island is the island of Corfu. This is a large island; its length is one hundred miles. There is a city which is well-populated, fertile and well defended and the city is situated on an impregnable pinnacle. The city's people are ready to defend themselves against those that attack them.

From Corfu to Othonoi is thirty miles and between Corfu and the city of Otranto already mentioned is ninety miles or a day by boat and likewise from Corfu to Vlorë is one day's sailing.

Of the islands in this area there is Cephalonia, to the east of Corfu and the circumference of Cephalonia is two hundred miles, the island is prosperous and there is a city there.[17] From Cephalonia to the island of Zakynthos is fifty miles and this is an inhabited island and its circumference is eighty miles.

From Cephalonia in a northward direction to the island of Lefkada is forty miles, this island is triangular in shape and each side is twenty miles in length. And we have now completed the description of this section thanks to Allah for that.

The Third Part from the Fourth Clime is now finished, thanks be to Allah. Now, we proceed to the Fourth part, Allah willing.

---

**15** Al-Idrisi says '*bi-l-rūmiyya*' meaning Greek; however, the medieval and modern Greek name for the city is *Avlónas*.

**16** Butrint was occupied by Robert Guiscard in 1081 to be used as a base for further attacks against the Byzantine empire. Richard Hodges, William Bowden, and Kosta Lako, *Byzantine Butrint: Excavations and Surveys 1994–99* (Oxford: Oxbow Books, 2020), 22.

**17** Robert Guiscard died from a fever in Cephalonia in 1085 along with 500 Norman knights. He had returned to Cephalonia and Zakynthos with 150 ships to retake the islands after they were lost to Byzantine forces by Robert's son, Bohemond. Graham A. Loud, *The Age of Robert Guiscard*, 223.

# CLIME FIVE, PART TWO (Partial)

[The first section deals with parts of France including Touluse and Burgundy and central and northern Italy including Venice, Genoa, Florence and Rome. We pick up the description from Rome to Reggio].

DESCRIPTION OF THE coastal route from Rome to the city of Reggio along the strait of the island of Sicily:

From Rome to Torre Astura is thirty miles and from Astura to Anzio is ten miles. Anzio is a secure port with deep water. And from Anzio to the mountain of Circeo[1] [*Jarnjwā*], also known as *Jarjīr* which is called *'al-Qīyiṭna'* of the Arabs, is thirty miles and this is a large river.[2] From there to the city of Terracina is six miles.

From the Rome River to Terracina is seventy-six miles. Terracina is a pleasant and fertile city, well-populated but its port is narrow and not of good quality. From Terracina to the city of Gaeta is twenty-four miles.

The city of Gaeta has a vast territory and many people. It is located on a promontory cut off from the land. Gaeta has a good port which is a safe haven in winter, the port is located on both the land and sea and soldiers are garrisoned there. Boats, both big and small, are constructed there.

From the city of Gaeta to Garigliano at the mouth of the river Sessa Aurunca is fifteen miles.[3] The Garigliano is a large, running and navigable river and on the river there are two towers. From there to the mouth of the Savone [*Jalāj*] River is twelve miles and this is located on an exposed beach[4] which cannot accommodate large boats. From there to the mouth of the Volturno [*Qabwa*] river is six miles and is also located on a beach without protection.

---

**1** Capo Circeo.

**2** The meaning of this word is ambiguous; Bresc and Nef translate it as *magasin* "warehouse/depot" and Amari and Schiaparelli as *Cala degli Arabi* "naval station of the Arabs." It derives, probably, from *qaṭṭana* "to live/settle/reside in/lodge in" giving the sense of a base of operations. It is possible this was a base for Aghlabid forces as, after years of piracy along the Tyrrhenian coast, by the ninth century the Aghlabids had become well entrenched in South Italian politics, establishing military and trade alliances with cities such as Naples, Gaeta, Salerno, Capua and Amalfi. Hilmar Krueger, "The Italian Cities and the Arabs before 1095," in *A History of the Crusades*, ed. Marshall W. Baldwin and Kenneth Meyer Setton, 6 vols., vol. 1, (Madison: University of Wisconsin Press, 1969), 40–54.

**3** Garigliano was the site of a Muslim settlement, one of only a few on the west coast of Italy, founded in 883 through negotiation with Docibilis I of Gaeta (d. 914) who needed Saracen military support against his rival, the count of Capua. The settlement lasted thirty years until it was given the coup de grâce by a Christian coalition which included Gaeta. Marco di Branco and Kordula Wolf, "Hindered Passages: The Failed Muslim Conquest of Southern Italy," *Journal of Transcultural Medieval Studies* 1 (2014): 51–74. See also Kreutz, *Before the Normans: Southern Italy in the Ninth and Tenth Centuries*, 62.

**4** *Iblāya*, this is an obscure word but may be related to Spanish *playa* (beach) from the Latin *plaga* "tract/region/quarter/zone."

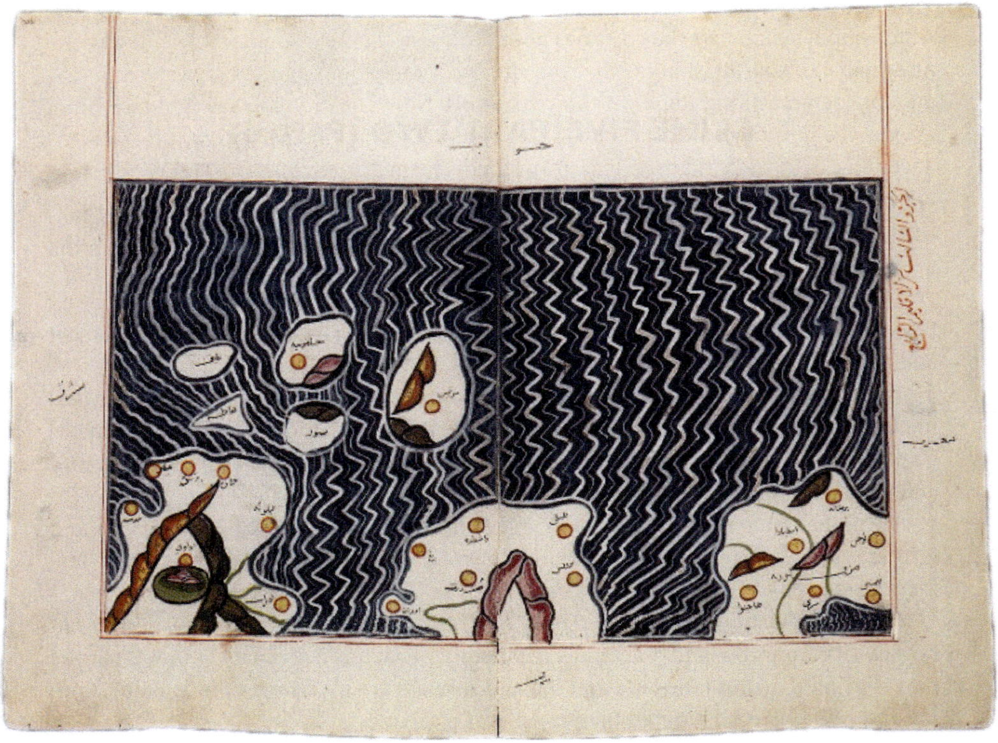

Figure 5. Heel of Italy and Mediterranean Islands (Oxford, Bodleian Library, MS Pococke 375, fols. 201b–202a). Reproduced under the terms of Creative Commons licence CC-BY-NC 4.0 with permission of the Bodleian Libraries, University of Oxford.

Map 4. Cities in South Italy. Map by authors.

From the Volturno River to Patria is twelve miles and this is a village and a port which is not secure. From Patria to the city of Cuma is six miles and this is a small city, a short distance from the sea. From Cuma to the port of Miseno is twelve miles and this is a secure port with little freshwater. Soldiers and men, I mean soldiers of both land and sea, are garrisoned there.

From Miseno to Pozzuoli [*Qashtlī*] is eight miles and this is an inhabited castle which resembles a small city, located on the edge of a bay. From Pozzuoli passing on to the city of 'Naples of linen' [*Nābil al-Kattān*] is twelve miles.[5]

'Naples of linen' is a pleasant, ancient and prosperous city. Its markets are full of sought-after goods and merchandise. From there to the port of Stabia is thirty miles and this is a good anchorage with plentiful water, it is located at the mouth of a freshwater river within a bay. And those wishing to reach Amalfi can do so walking overland at a distance of fifteen miles.

Between Stabia and Naples there is the mountain Vesuvius [*Jabal al-Nār*] and this is an area that you cannot reach as the volcano continually spews forth fire and rocks.[6] And for those wishing to travel along the coast from the city of Stabia to the city of Sorrento the distance is thirty miles.

The city of Sorrento is located on a promontory and is pleasant and prosperous with beautiful homes, abundant products and trees. There is a barely accessible canal in which boats cannot anchor in winter but rather are tugged in. Sorrento has facilities for boat building.

From the city of Sorrento to Punta Campanella [*Rās Mantīra*] is twelve miles and from there to Positano is fifteen miles. Positano is a small port and from there to the city of Amalfi along the coast is eighteen miles. Amalfi is a prosperous city which is well-fortified on the landward side but easy to take from the sea and those who attack it will take it. Amalfi is an ancient city and its people are very well to do.

From the city of Amalfi to the mouth of the Vietri [*Bādhrwā*] River is ten miles and this is a good place for fleets to dock. At the top of this river is a resting place called Vietri and the river is named after it. Vietri is a fortified locality not reachable except for entry via two gates, there is water and wood there. From this river to Salerno is two miles.

Salerno is an impressive city with plentiful markets, public utilities, wheat and grain. From Salerno to the Asa River, which is a narrow anchorage, is six miles and from there to the Sele is twelve miles. The Sele River has abundant water and is navigable, its banks are forested and swampy, rendering them difficult to access, therefore

---

**5** This is in keeping with Ibn Hawqal's tenth century description of Naples: "the main wealth of Naples is linen and linen cloth. I have seen there pieces the like of which I have seen in no other country...They are woven 100 dhira' by 15 or 10, and they sell for 150 ruba'i [37.5 dinars]." Robert S. Lopez and Irving W. Raymond, *Medieval Trade in the Mediterranean World* (New York: Columbia University Press, 1955), 54. A *dhira'* is approximately half a metre while the *ruba'i* was a quarter-dinar. See also Paul Arthur, "Naples: A Case of Urban Survival in the Early Middle Ages?," *Mélanges de l'école Française de Rome* 103 (1991): 774; Michele Amari, *Storia dei Musulmani di Sicilia*, vol. 2 (Florence: Felice le Monnier, 1858), 178.

**6** This is literally "The Mountain of Fire." This is the same name al-Idrisi uses for Etna in Sicily.

boats and fleets can dock there in safety. From there to the gulf of Agropoli and the island of Licosa close to the mainland is twenty miles, and there is no port there.

From the island of Licosa to The Gulf of the Two Rivers [*Jūn Al-Wādyīn*][7] is twenty miles and from there to Castellamare di Velia is ten miles.[8] From Castellamare di Velia to Molva is thirteen miles and the mouth of the Mingardo [*Shant Sīmr ī*] River is here, flowing past Molva. From there to Policastro is twenty-four miles and this is a large and prosperous castle and on its northern side is a river. From there to Petrosa, also known as the cape of Policastro, is six miles.

From this cape to Castrocucco is thirteen miles and from Castrocucco to Scalea is twelve miles. Scalea is a large castle and from there to the cape of Cirella [*Rās Jarla*] is nine miles. Cape Cirella is also a river and is navigable by lightweight boats.

From Cirella to Amantea is thirty-eight miles. Amantea is a pleasant and civilized city on the gulf of Oliva[9] [*Awbla*] and from Amantea to Sant'Eufemia is fourteen miles and from there to the Amato [*Qazlīt*] River ___[10] and from there to the Mucata River is two miles.

From the Mucata River to Francavilla Angitola which is a large, inhabited castle is thirteen miles. From Francavilla Angitola to Bivona is twelve miles and from there to Tropea is twelve miles. From Tropea to Capo Vaticano also known as *Qāmū* with 'mim' is six miles.

And so the total distance from the city of Amantea to Capo Vaticano is sixty-five miles.

From Capo Vaticano to Reggio is sixty miles and from Capo Vaticano to Vaticano is six miles. From Capo Vaticano to Tropea is six miles. Tropea is a pleasant city, and famous amongst the metropolises of the lands of Rum.[11] From Tropea to Nicotera is twelve miles and from *Al-Farū* to Reggio is twelve miles.[12]

And we will describe this region and its inland territories in part three.

And it remains now to describe cities beyond the city of Salerno. From these cities are the cities of Benevento and Avellino. From the city of Salerno to the city of Avellino is twenty-four miles in a northerly direction and from Benevento to Salerno is sixty miles and from Avellino to Cimitile is twenty miles and from Cimitile to Salerno is thirty miles.

The city of Benevento is an ancient and prosperous city and Avellino is a small city like a castle. From Benevento in a westerly direction to the city of Montesarchio is eighteen miles and from Montesarchio to Arienzo thirty-two miles.

---

**7** This is literally "The Gulf of the Two Rivers/Valleys." Location uncertain.

**8** Also known as Castellammare della Velia.

**9** The Oliva is a river which descends to Amantea.

**10** Distance missing from all manuscripts.

**11** Here al-Idrisi is most probably talking about "Christian" in a general sense. Tropea was a pontifical *massa* overseen by both Rome and Constantinople. Ghislaine Noyé, "Byzantine Calabria," in *A Companion to Byzantine Italy*, ed. Salvatore Cosentino (Leiden: Brill, 2021), 434, 440.

**12** Derived from the Latin *pharus*. The location of *Al-Farū* is uncertain.

Arienzo is a pleasant city, prosperous and in good condition. From Arienzo to Capua is thirty miles and Capua is situated on a large river which descends from the mountains in the vicinity of Benevento. From the city of Capua to Aversa is eight miles and from Aversa to Naples is twelve miles and we have now included all that is in this part, thanks be to Allah.

Part Two of Clime Five is finished and will be followed by Part Three, Allah willing.

Figure 6. Gulf of Venice (Oxford, Bodleian Library, MS Pococke 375, fols. 250b–251a). Reproduced under the terms of Creative Commons licence CC-BY-NC 4.0 with permission of the Bodleian Libraries, University of Oxford.

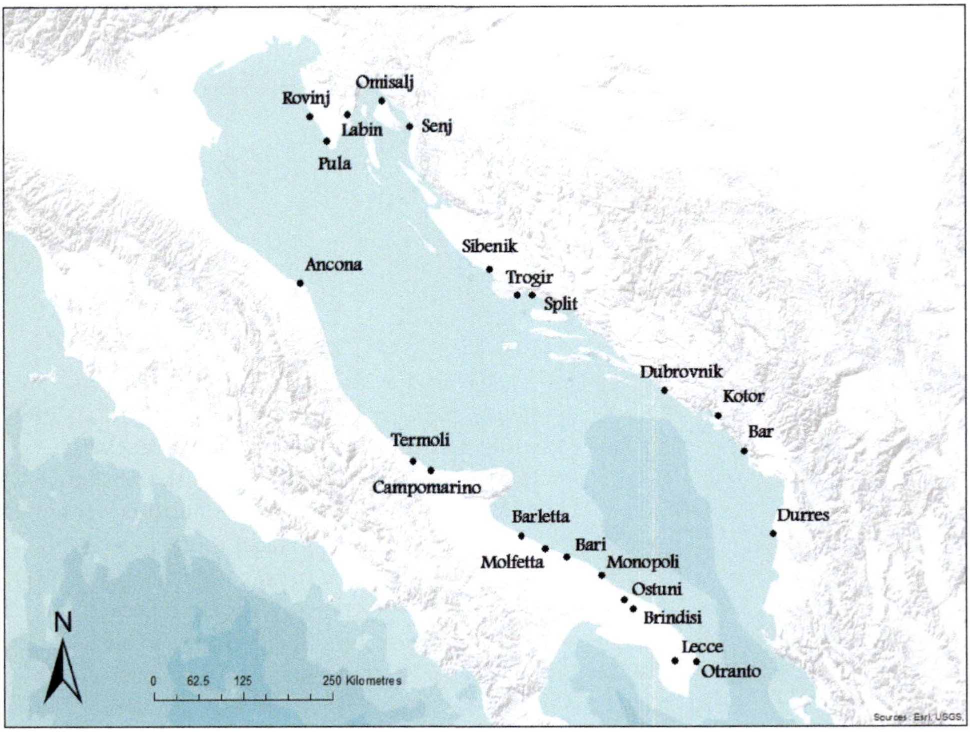


Map 5. Cities in the Adriatic. Map by authors.

# CLIME FIVE, PART THREE

THIS THIRD PART of the Fifth Clime includes part of the lands of Calabria and the lands of the Lombards and the major part of the Adriatic Sea and the well-renowned lands within it.

On the eastern shore of the Adriatic are the cities of Rovinj, Pula, Labin [*Darūna*], Senj [*Āsyā*], Omišalj [*Miṣqila*], *Arns*,[1] Zaton, Nona, Zadar, Šibenik, Dubrovnik [*Raghūṣ*], Split, Trogir, Kotor, Bar [*Antabariyya*], Đurinići, Durrës, Bitola, Vonitsa, *Fāmīyu* and *Kīra*.[2]

And the towns on the western side of the Adriatic are Brindisi, Ostuni, Monopoli, Conversano, Molfetta, Biscgelie, Trani, Barletta, Cannae, Siponto which is also called Vieste, Rodi, Lesina which is also called *Lāzna* and Campomarino. All these places are part of the countries of the Lombards on the western side of the Adriatic Sea.

Other coastal places include Termoli, Tronto, Numana and Ancona. And on the Mediterranean coast there is Tacina, Crotone, Rossano, Roseto and Taranto.

In the countries of Calabria there are Catanzaro, Martirano, Viggiano, Castrovillari, Benevento, inland Melfi, Conza, Venosa, Sant'Agata de'Goti, Chiaromonte, Senise, Bisignano, Simeri, Strongoli, Tricarico, Acerenza and all these places are within the countries of Calabria.[3]

And among the Lombard regions there are Cerignola, Matera, also known as *Mātla*, Cerignola, Mottola, also called *Māt ī*, Gravina, Canosa, Ordona, Ascoli Satriano also called *ʾAsqala* with an 'sin', San Lorenzo, Sambiase, Civitate, San Severo, Monte Sant'Angelo, Lesina, Campomarino and Termoli.

Now we will describe place by place and we will present descriptions and itineraries just as we have done for the previous Climes, by the will of Allah almighty. And we will start with a description of the route from the city of Reggio along the Mediterranean coast and then along the Adriatic coast to the city Ancona.

From Otranto to the cape of Roca Vecchia [*Shūdha*] is twelve miles and from there to *Shant Juwān Martūblī* is twelve miles and this is a pleasant village that resembles a city and from there to Torre Chianca is six miles.[4]

From Torre Chianca to Torre San Gennaro is twelve miles and from there to the cape of Torre Cavallo [*Nāwra*] is twelve miles and from there to the city of Brindisi is four miles. So, from Otranto to Brindisi travelling along the various gulfs is a total of fifty-eight miles and in a direct line, forty-eight miles.

Brindisi is a splendid city and is surrounded by the sea on three sides, in similar fashion to Constantinople the Great. The city is well-constructed with spacious squares; it is blessed in its fertility and there are many amenities. From Brindisi to

---

**1** The location of *Arns* is unknown.

**2** The location of these two places is unknown.

**3** Al-Idrisi uses "inland Melfi" to avoid confusion with coastal Amalfi.

**4** The location of *Shant Juwān Martūbly* is unknown.

Guaceto is twelve miles, Guaceto is made up of three pleasant islands and between them and the mainland is half a mile.

From Guaceto to the port of Saint Nicholas is twelve miles and this is a secure port with freshwater.[5] From there to the city of Monopoli is twenty-four miles. Monopoli is a small and civilized city and from there to the castle of Polignano is six miles, near to Polignano there is a city called Conversano and this city is nine miles from the sea.

From Polignano to the port of San Vito is two miles and from San Vito to the city of Bari is twenty-two miles. Bari is a large and well-populated city situated at the bottom of a gulf. Bari is the principal city of the Lombards and contains an arsenal and it is one of the most famous capitals of the Rum.[6]

From Bari to Torre Santo Spirito, the tower of *Ajīlūwā*, also known as *Ishīlūwā*, which connects inland with the city of Bitonto. Between Bitonto and the sea is six miles. From there to Giovinazzo is six miles and then to Molfetta is four miles and Molfetta is also known as *Malfant* with the letter *fa*. Inland from Molfetta is Ruvo, Ruvo is a pleasant city of medium size and between it and the sea is six miles. From Molfetta to Bisceglie is ___.[7] Bisceglie connects directly inland with Corato and between Corato and the sea is nine miles.

Corato is a pleasant, civilized and noble city; it has a vast territory, fertile with many fruits and a good standard of living. From Corato to Trani on the coast is eight miles and likewise from Bisceglie which has already been mentioned to Trani is six miles.

Trani is a medium-sized city with a wall and a well-known market. From Trani to Barletta is six miles along the coast. Connecting with Barletta there is a city called Andria, situated on a plain, which is large and well-populated and between there and the sea is nine miles.

From Barletta along the coast to the Ofanto River [*Wādi Lūdhra*] is six miles. On this river there is a large monastery called Santa Maria and from the monastery to the city of Cannae which is higher than sea level is four miles. Cannae is a small and civilized city with much trade and wealth and its people are well to do.

---

**5** Location unknown, though Amari and Schiaparelli suggest Torre Pozzelli. Al-Idrisi, *L'Italia descritta nel "Libro del Re Ruggero"*, 103. Saint Nicholas of Myra (270–343 CE) is associated with the city Bari; his bones were brought to the city from Asia Minor in the twelfth century by a group of local sailors but were later taken to Venice. Battista Bronzini, "Culto e leggenda di san Nicola (quattro note storico-antropologiche)," *Lares* 68 (2002): 5–19; Dawn Marie Hayes, "The Cult of St Nicholas of Myra in Norman Bari, c. 1071–c. 1111," *The Journal of Ecclesiastical History* 67 (2016): 492–512.

**6** Al-Idrisi most probably means "Christians" in a general sense here, both Orthodox and Latin, as Bari was variously under Lombard, Byzantine, and Norman control (as well as Muslim from 847–871). Indeed Bari became the site of an attempt to reconcile Orthodox and Latin Christianity when in 1098 Pope Urban II convened the Council of Bari, an attempt to heal the schism by addressing theological disagreements between the Churches. Dawn Marie Hayes, The cult of St Nicholas of Myra in Norman Bari, ca. 1071–ca. 1111, 506. Bari was also the capital of the Byzantine catepanate of Italy.

**7** Distance missing from all manuscripts.

From Santa Maria to Torre delle Pietre Rosse [*Shant Niqūla Bibitra*][8] is twelve miles and this is located on the sea on a promontory connected inland with the city of Salapia which is six miles from the sea.[9] From Salapia to the Rivolo River, also known as *Nīqlū* is twelve miles and the actual name of this river is Candelaro.[10]

From the Candelaro River to the Cervaro River is eleven miles and from there to the city of Siponto is two miles, Siponto is close to the sea. From Siponto to Mattinata is twelve miles and Mattinata is also close to the sea. From Mattinata to Monte Sant'Angelo which is above sea level is eight miles.

And also from Mattinata to the port of San Felice is twelve miles and this is a village with a great church. From San Felice to Vieste [*Bistiyya*] which lies at the bottom of the gulf on a promontory overlooking the sea is twelve miles. The distance from the village to the head of the promontory is the length of an arrow shot and the width of the promontory at its head is half a mile and the width at the promontory's mid-way point is four miles.

From Vieste to Peschici is twelve miles and from Peschici to Rodi Garganico is eight miles. From Rodi Garganico to Cagnano Varano is twelve miles and from there to Devia is eleven miles. From Devia to Lesina is eight miles and from the sea to *Aqrabīl* is four miles, Lesina is close to the sea and is also known as *Lāzna*. [11]

From Lesina to Campomarino is twelve miles and from Campomarino to Termoli is twenty miles and this is also known as *Tarmalis*. Between Lesina to Termoli there is a curved beach. From Termoli to the mouth of the Pescara River is fifty-nine miles in a southerly direction. Like this from the Pescara River to the Tronto also known as *Trant* it is thirty-six miles. The Tronto River is large and located along it, far from the sea is the city of Tronto.

Tronto is a large city which abounds with fertility and bounties. From the mouth of the Tronto River to the city of Numana on the coast is fifty-eight miles. Numana is a big city with a large territory and plentiful gardens and grapevines and from there to the city of Ancona is six miles.

Ancona is an ancient and famous city, one of the main centres of the lands of the Rum[12] who live on the Adriatic Sea and we have mentioned Ancona previously.

---

**8** This is literally "Saint Nicholas in Stone" (an Arabized form of *petra*).

**9** Salapia (also known as Salpi). Frederick II (1194–1250) constructed a major hunting residence here. Pietro i Biase, *Puglia medievale e insediamenti scomparsi: la vicenda di Salpi* (Fasano: Schena Editore, 1985); Giovanni De Venuto et al., "Salapia: Storia e archeologia di una città tra mare e laguna," *Mélanges de l'École française de Rome—Antiquité* 127, no. 1 (2015), https://doi.org/10.4000/mefra.2719, accessed November 21, 2023.

**10** The author has confused the Carapelle with the Candelaro river according to Amari and Schiaparelli. Al-Idrisi, *L'Italia descritta nel "Libro del Re Ruggero"*, 104n6.

**11** The location of *Aqrabīl* is unknown.

**12** Here al-Idrisi intends Byzantine; Ancona was an important Byzantine stronghold during the Gothic war of the sixth century. Vera Von Falkenhausen, "Ancona," in *The Oxford Dictionary of Byzantium*, ed. Alexander P. Kazhdan (Oxford: Oxford University Press, 1991); David Abulafia, "Ancona, Byzantium and the Adriatic, 1155–1173," *Papers of the British School at Rome* 52 (1984): 195–216.

Between the city of Ancona and Campomarino is an empty wilderness which takes twelve days or three hundred miles to traverse. And in this wilderness there are people who take refuge in the woods where they hunt and collect honey. We have already described in the previous section the area from Ancona to the edge of the gulf [of Venice].

From there, the bay turns eastward and this is the land of Aquileia and among the inland towns of Aquileia is Vranja also known as *Barāna* as well as Buje and Materada [*Ṭāmṭaras*]. The city of Vranja is large and between it and the city of Materada is one day's light travel and from Materada to the city of Buje is nine miles. Buje is a large and prosperous city and from there to Izola also known as *Injala* whose population is Frankish, is three miles. From Izola to Frankish *Qandīla* is three miles and from there to Vranja is two miles and we have already mentioned this place.[13]

These are all the inland towns of Aquileia. The coastal towns of the region include Koper [*Dastarīs*] and between there and Materada, which is a city of Aquileia, is twenty-three miles. From Materada to Umag also known as *Awmāghu* is nine miles and like this from the inland city of Lovran to Umag is eighteen miles and its people are Frankish. Umag is on the coast and from there to the city of Novrigrad [*Jabṭanūba*] which is a new Frankish town and is really two cities in one: one is on the coastal plain and the other is located on a mountain overlooking the sea at a distance of eight miles.[14]

And from Novrigrad to Poreč [*Baranjwā*], also known as *Barnazwā*, is twelve miles. Poreč is a flourishing and prosperous city and it has a fleet with many boats. From Poreč to Rovinj which is in the hands of the Franks is fifteen miles and Rovinj is a pleasant city with a large territory and much prosperity. From there to the city of Pula is twelve miles, Pula has a large territory and much prosperity and its fleet is always ready for action. From Pula to the prosperous city of Medulin is sixteen miles and from there to Labin is forty miles. From Labin to Plomin [*Filāmūna*] is six miles and both these cities are prosperous, their territory is contiguous and their conditions similar.

From Plomin to Lovran is four miles and the city of Lovran is very prosperous with excellent conditions, boats that are ready for action and a permanent arsenal; Lovran is the last town in the land of coastal Aquileia. Adjacent to this region in an easterly direction are continuous mountain ranges and wilderness cut off from development. And following the cities we have already mentioned in the land of Aquileia, are the lands of Croatia which is called Dalmatia and these are on the coast. The first town in the lands of Croatia is the city called Bakar and from there to Lovran which we have already mentioned is ten miles. Bakar is a pleasant, civilized city and it is six miles from there to the city of Lopar which is a large and prosperous city on the slope of a mountain.

---

**13** The location of *Qandīla* is uncertain.

**14** Novrigrad was previously Cittanova.

From Lopar to the city of Senj is thirty miles. Senj is a pleasant and prosperous city, its people are Slavic and have many boats. From Senj to Baška is fifteen miles and its people are Slavic, it is a small city with a small number of boats. From Baška to the city of Omišalj [*Maṣqala*] is twenty miles and this is a medium-sized city which belongs to the Dalmatians, and from there to *Arnaṣ* is fifteen miles and this belongs to the Dalmatians, and they have a number of boats, and from there to the city of Zaton is thirty miles.[15]

Zaton also belongs to the Dalmatians and they have war ships.[16] From Zaton to the city of Nin, also known as *Nīnṣ*, is twenty miles and Nin is a large and attractive city in a secure position. From there to Zadar,[17] Zadar is a large city with extensive cultivation and widespread vineyards and it is on the coast, its people are Dalmatian and the sea crashes against its city wall.

And from Zadar to Biograd [*Daghwāṭa*] is thirty miles and this is a city whose people are Dalmatian and Slavic. It is one of the capitals of the Rum and its people are brave.[18] From Biograd to Šibenik is twenty miles and this is a pleasant city with much cultivation and traders travel to and from the city by land and sea.

From Šibenik to the city of Rogoznica [*Wawghūry*] is fifty miles and this pleasant city is also known as *Lūghārū* and is one of the most attractive and secure capitals. Its people are Dalmatians and they are travellers and wanderers and they have war ships.

From Rogoznica to the city of Trogir also known as *Targhūra* is six miles and its people are Dalmatians, they are boat builders, warriors and travellers.

And from the city of Trogir to the city of Split is twelve miles and the people of Split are Dalmatians and this is a prosperous city with a large territory with much cultivation and commerce. The entire city is paved and they have war ships.

From Split to the city of Ston [*Shathghanū*] is twenty-five miles and its people are Slavic, they have a wide territory, expansive cultivation and many boats.

And from there to Dubrovnik which is also called *Raghūṣa* is thirty miles, its people are Dalmatians, they have war ships, they are brave and steadfast and this is the last town in the region of Croatia.

From Dubrovnik to the city of Kotor, also known as *Qādhruwā*, is twenty miles and this is a pleasant and populous city; its people are Dalmatians and they are invaders and travellers and they have a number of boats.

From Kotor to Bar [*Antābrwā*] is thirty miles and its people are Slavic. Bar is a large, prosperous city, one of the most famous capitals and from there to Ulcinj [*Dhalūjiyya*] *is seventy miles and this is one of the capitals of Slavonia.*

**15** The location of *Arnaṣ* is uncertain.

**16** This is literally *marākib ghazwāniyya* from *ghazū* meaning "assault/raid/incursion/inroad/invasion/attack."

**17** Distance missing from all MSS.

**18** In this case, probably Latin Christians; Biograd was the seat of the Roman Catholic Diocese of Alba Maritima (also known as the Diocese of Zaravecchia) from the mid-eleventh century. Florin Curta, *Southeastern Europe in the Middle Ages, 500–1250* (Cambridge: Cambridge University Press, 2006), 266.

*From Ulcinj to the city of Durrës which belongs to the Franks is eighty miles.*[19] Durrës is the closest point to Otranto on the other side of the strait and between the two is senventy miles.

From Durrës to Himarë is one hundred and twenty-five miles, ascending the wall.[20] Himarë's old name was *Barnatū*. And this is the extent of the Gulf of Venice, its towns and strongholds and we have presented a sufficient amount [of information] for those interested and curious, and thanks be to Allah for that.

As for the islands that are in this sea [the Gulf of Venice], we wish to mention them in succession until we fulfill the purpose that we intended as we described it. In this Venetian Sea is the island of Losinj [*Awser*] which is close to the mainland from one of its points, the rest of the island descending towards the sea, and between the two is eight miles. Losinj is in the gulf of Istria and its length is twenty miles, its width is twelve miles and it is prosperous.

And from there to the island called Cres [*Jarsa*] is five miles and between this island and the mainland is six miles. This is a large and very prosperous island and its length is more than its width, its length being around sixty miles and its width twenty-five miles. A count and a bishop reside there.

From Cres to the island of Rab is six miles and it is opposite the Croatian mountains and between the island and the mainland is twelve miles. Its length is around thirty miles and its width eighteen miles and a count and a bishop reside there.

From Rab to the island of Pag is four miles and Pag is opposite Nin and between Pag and the mainland is four miles and its length is about twenty miles and its width ten miles.

All the aforementioned islands are prosperous, and they are part of the region of Croatia. As for the islands of the Venetians, there are six, three in a row, and three in another line, and all of them are inhabited.

And in the strait we have already described there is an uninhabited island which is close to Vlorë.[21] Likewise close to Brindisi there are three islands called Torre Guaceto and they are close to the mainland, very small and uninhabited. And this is all the islands found in this sea, both inhabited and uninhabited.

---

**19** Al-Idrisi uses the term Franks, a generic term for the Latin Christians of Europe, but almost certainly means the Normans. The Normans attacked the city several times: Robert Guiscard took the city in 1081, Bohemund besieged it in 1107–08; in 1185 William II of Sicily pillaged it; however, throughout the first half of the twelfth century (when Al-Idrisi was composing the book), Durrës was largely in Byzantine hands, the city playing an important part in Basil II's war against Bulgaria. Timothy E. Gregory, "Dyrrachion," in *The Oxford Dictionary of Byzantium*, ed. Alexander P. Kazhdan (Oxford: Oxford University Press, 1991); John Van Antwerp Fine, *The Early Medieval Balkans: A Critical Survey from the Sixth to the Late Twelfth Century* (Ann Arbor: University of Michigan Press, 1991).

**20** Al-Idrisi is probably talking about a land route which involves following the mountain ranges southward for a period.

**21** Sazan.

We will return now and describe the mainland [of Italy], its configuration, its routes and distances, the conditions of its inhabitants and the position of its walls, city by city and region by region.

And we will describe the route from Taranto previously mentioned to Naples. From Taranto to Matera is sixty miles and from Matera to Gravina is sixty miles. From Gravina to Canosa is one hundred and twenty miles and from Canosa to Andria is eighteen miles.[22] From Andria to Trani is eighteen miles then to *Bābra* is fifteen miles.[23] From *Bābra* to Frigento is twenty-six miles then to Cimitile is eighteen miles and from there to Naples on the coast is thirty miles.

And all these aforementioned cities and famous and invincible strongholds are similar in that they are metropolises and share similar conditions. They are prosperous by all types of trade, fertility abounds and security is stable and most of this territory, or rather all of it, is part of the regions of Calabria and Apulia.

Apulia and Calabria are the name of the lands and the regions which are made up of many cities and the first among these is Reggio situated on the straits of Sicily, Reggio is small and civilized and has markets and buildings and is a meeting place for people who come and go.

From Reggio to Tuccio is one day's travel and from there to Gerace is one day's travel too. From Gerace to *Almāṣa* is seventy-six miles.[24] *Almāṣa* is a small city with a market and fruits and from there to Le Castella is thirty miles, and this is a small city.

From Le Castella to Crotone travelling directly is thirteen miles and following the curvature of the coast is eighteen miles. From Crotone, for those who want to cut across the sea is one day's sailing plus thirty miles, and we have described the details of this place previously.[25]

And from the city of Gerace to Stilo is twenty-four miles. Stilo is small but well-populated with many advantages and from here to Catanzaro is twelve miles. Catanzaro is a secure castle and from here in a westerly direction to Sant'Eufamia which is on the sea is twelve miles and Sant'Eufamia was mentioned previously, all these places are within the territory of Calabria.

From Crotone along the coast to Tacina on the sea is three Frankish miles and this is nine miles.[26]

From Crotone to Belcastro [*Jinqū Qasṭrūwā*] is three Frankish miles and likewise from Tacina to Belcastro.[27] From Belcastro to Simeri is fifteen miles and between Simeri and the sea is three miles. All these towns are small though civilized with markets, buying and selling and they are similar to each other in their characteristics and

**22** Al-Idrisi appears to have confused Canosa and Venosa.

**23** The location of *Bābra* is uncertain. Bresc and Nef suggest Rapolla though this is around 100 kms from Trani. Al-Idrisi, *La Première géographie de l'Occident*, 389.

**24** Possibly Masa, see Clime 4, Part 2 note 9.

**25** To the opposite side of the Gulf of Taranto.

**26** The Tacina River whose source is in Sila, a mountainous plateau covering the provinces of Cosenza, Crotone and Catanzaro.

**27** Distance missing from all MSS.

conditions. From Simeri to Catanzaro is fifteen miles and also from Simeri to Taverna is eighteen miles and from Simeri to Strongoli is twenty-one miles.

From Strongoli to Crotone is twenty-four miles and between Strongoli and the sea is six miles and also from Strongoli to Umbriatico is eleven miles and from Umbriatico to Pietrapaola is twenty-seven miles.

From Pietrapaola to Cirò [*Ibsakhrū*] is thirty-three miles and between Cirò and Rossano on the coast is fifteen miles.[28] From Rossano to San Mauro is five miles and between San Mauro and the sea is six miles and from San Mauro to Armento is three miles. From Armento to Sant'Arcangelo is six miles.

And from San Mauro to Bisignano is nine miles and Bisignano to Acri is twelve miles and from Acri to Sant'Arcangelo is twelve miles and from there to Rocca Nova [*Ruqqa Fīlib*] is six miles. From Sant'Arcangelo to Colobraro is twelve miles and from Sant'Arcangelo to Senise on the right is twelve miles and from Sant'Arcangelo to Ganano on the left is twelve miles.

And the Acri river separates these two places. From Sant'Arcangelo to Castel Missanello is six miles and this is a secure castle. From Castel Missanello to Gallicchio is two miles and from Sant'Arcangelo to San Brancato [*Bins Adrāt*] in a westerly direction is six miles and from San Brancato to *Qishtāl Lūrint* is six miles and from there to the city of San Martino di Agri is three miles.[29]

From San Martino di Agri to Montemurro is six miles and from Montemurro to Viggiano is six miles and from Viggiano to Marsicovetere is six miles and from there to Grumento Nova [*Ṣabūnāra*] is twelve miles and from there to Sarconi is three miles and from Senise to Tursi is twelve miles.[30]

From Tursi to Sant'Arcangelo is twelve miles and from Sant'Arcangelo to the castle of Anglona is six miles.[31] From Anglona to Castel di Sangro is twelve miles and from Castel di Sangro to Taranto it is forty-eight miles.[32]

We return now and say that from Senise to Tursi is twelve miles and from Tursi to Anglona is six miles and from Anglona to the castle of Sant'Arcangelo is eighteen miles, as we mentioned previously.

---

**28** Cirò is ancient Psicro.

**29** The location of *Qishtāl Lūrint* is uncertain though it is an Arabized version of "The Castle of Lorenzo." Amari and Schiaparelli place this at "fiume di Lorenzo" which empties into the Agri river. Al-Idrisi, *L'Italia descritta nel "Libro del Re Ruggero"*, 113n5. Bresc and Nef place the source between Aianello and Misanello. Al-Idrisi, *La Première géographie de l'Occident*, 390n2.

**30** Known as Saponara di Grumento until 1932 and site of the ancient Roman city of Grumentum. The original village was conquered by Robert Guiscard in the 1050s, he subsequently constructed a residence there, and a village developed around it. A wall with seven gates was then built; the entire village, thus fortified, took the name of Saponaria. The name appears for the first time in the Bull of Gisulfo of 1095. Vincenzo Falasca, *Il castello Sanseverino di Grumento Nova* (Grumento Nova: Patrocinio Amministrazione Communale, 2010); Gabriele De Rosa and Antonio Cestaro, *Storia della Basilicata, Vol. 2, Il Medioevo* (Bari: Editori Laterza, 2021).

**31** Ancient Anglonum.

**32** Castel di Sangro is in Abruzzo, meaning al-Idrisi has made a substantial error here.

And from Sant'Arcangelo to Carbone is eighteen miles and from Carbone to the castle of Colobraro is twenty-four miles. The castle of Colobraro is situated parallel with Carbone, positioned behind the mountain, and Carbone is also parallel with the city of Calvera and between the two is six miles.[33] Between these two towns there is a region called Castronuovo di Sant'Andrea and between there and Calvera is three miles.

And from Castronuovo di Sant'Andrea to the city of Battifarano is three miles and between Battifarano and Senise is six miles and these four towns are behind the mountain.[34] From Sant'Arcangelo to Rocca Nova is six miles.

We return now and say that from Taranto in a southerly direction to the city of Gallipoli on the sea is sixty miles and from Gallipoli to Otranto in an easterly direction is three hundred and thirty miles. From Otranto to Lecce is seventy-two miles, Lecce is an inland city. And from the city of Lecce to Brindisi, situated on the Adriatic Sea and which we mentioned previously, it is seventy-two miles.

From Taranto to Pennapiedimonte is eighteen miles and from there to Forca is eighteen miles and from Forca to Castelnuovo[35] is twenty-four miles and from the castle of Pennapiedimonte to the castle of Civitaluparella is twenty-four miles and from there to the castle of Monteferrante travelling in an indirect line is fifteen miles.

From Monteferrante to *Ṣanāla* is twenty-four miles.[36] And from the castle of *Ṣanāla* to the city of Castel di Sangro is twenty-seven miles and likewise from the castle of Civitaluparella to *Ṣanāla* is twenty-four miles.

And we say that from the castle of Pennapiedimonte to the castle of Lama dei Peligni is twenty-one miles and from Lama dei Peligni to Carceri is twenty-seven miles and from there to the city of Castel di Sangro is fifteen miles. Likewise from Lama dei Peligni to the castle of Civitaluparella is fifteen miles and from the castle of Lama dei Peligni to the city of Taranto is six miles and from Taranto to the castle of Pacentro is eighteen miles.

From Pacentro to the castle of Forca is twelve miles and from Forca to Carceri is twelve miles and from the castle of Carceri to the city of Castel di Sangro is twelve miles and from Pacentro to Gagliano Aterno which is a pleasant castle, is eighteen miles.[37] From the castle of Gagliano Aterno to the castle of Valleoscura is fifteen miles

**33** Which mountain al-Idrisi is referring to is unclear but between Colobraro and Carbone there is a significantly mountainous area which is now the Parco Nazionale del Pollino.

**34** No city remains: Battifarano or Battafirano is now a vineyard and agriturismo business which has been run by the Battifarano family since the sixteenth century. Annunziata Bozza, ed., *Inventario dell'archivio privato Battifarano* (Basilicata: Dipartimento Formazione Lavoro, Cultura e Sport, 2009), www.regione.basilicata.it/giunta/files/docs/DOCUMENT_FILE_324272.pdf, accessed November 21, 2023.

**35** Amari and Schiaparelli state this is Castel Nuovo though which one they are referring to is unclear, most probably Castelnuovo (Avezzano) which fits in with al-Idrisi's itinerary and distances and was occupied by the Normans. Al-Idrisi, *L'Italia descritta nel "Libro del Re Ruggero"*, 115.

**36** Amari and Schiaparelli identify this as Sonella. Al-Idrisi, *L'Italia descritta nel "Libro del Re Ruggero"*, 115n4.

**37** Ancient Balianum.

and from Vasto to the city of Castel di Sangro is twelve miles and also from Valleoscura to the castle of Pacile is eighteen miles in a westerly direction.

And from Vasto, and this is the rock we have already mentioned, to Castelnuovo is six miles and from there to Pescasseroli is eighteen Frankish miles which is twenty-four miles and this is a village which resembles a city. From Pescasseroli to the castle of San Donato Val di Comino is twenty-four miles and from there to *Rālbinū* is eighteen miles.[38]

From *Rālbinū* to Cerro which is a secure castle is twenty-five miles and also from *Rālbinū* to the city of Castel di Sangro is eighteen miles and from the castle of Castro [*Qāsṭrwā*] to the castle of Villetta Barrea is twenty-four miles. And from Villetta Barrea to Castel di Sangro is eighteen miles and also from San Donato Val di Comino to *Ruqqa Shanj* is eighteen miles and from there to Alfedena is three miles.[39]

From Alfedena to the city of Castel di Sangro is twelve miles and from Rocca Giuliopoli to Alfedena is eighteen miles and from Cerro to Atina is twelve miles and from Atina to Castel di Sangro is eighteen miles.

Now we return and say that from Castel di Sangro to the castle of Agnone is thirty-six miles and from there to Castel di Jonato [*Makshatr Yināṭa*] on the right is nine miles and on the left is Pietrabbondante which is fifteen miles from the castle of Palena, passing Castel di Sangro to the north and slightly to the right is the castle of Pacentro and between the two places is eighteen miles. And from Palena to Castel di Sangro is thirty-six miles.

From the castle of Cerro mentioned previously to Fallo is twenty-four miles and from there to San Donato Val di Comino is eighteen miles and from there to Scanno is twenty-four miles and from there to Valleoscura is twenty-one miles. From Valleoscura to Campo di Giove is eighteen miles and from there to Lama dei Peligni is twenty-four miles and from there to the city of Taranto is six miles.

From Lama dei Peligni to Pennapiedimonte is eighteen miles and from the castle of Pennapiedimonte to Colle Archiano is twenty-four miles and from Archiano to the castle of Carunchio is twenty-four miles and also from Archiano to the castle of Montebello sul Sangro is eighteen miles and from Montebello sul Sangro to Lama dei Peligni indirectly is twenty-four miles.[40]

And from Fallo to the castle of Montemitro eighteen miles and from *Bāṭra*[41] to Castel del Giudice is fifteen miles then to Montemitro is fifteen miles and from Calcasacco

**38** Amari and Schiaparelli identify *Rālbinū* as Rocca Albano. Al-Idrisi, *L'Italia descritta nel "Libro del Re Ruggero"*, 116n6.

**39** Amari and Schiaparelli identify *Ruqqa Shanj* as Rocca Sangia. Al-Idrisi, *L'Italia descritta nel "Libro del Re Ruggero"*, 117n1.

**40** Colle Archiano is also known as Casale Archiano, located south of Tornareccio, near the hamlet of S. Giovanni. Carlo Tedeschi, "Monasteri Dell'Abruzzo Adriatico: Un Dossier Documentario (1019–1065)," *Mélanges de l'École Française de Rome—Moyen Âge* 128, no. 2 (2016), https://doi.org/10.4000/mefrm.3407, accessed November 21, 2023.

**41** Given the distances, al-Idrisi is presumably referring to Pietrabbondante, mentioned previously, though in this case he uses *ṭa* rather than *ta*.

to the castle of Castiglione Messer Marino is eighteen miles and from there to Castel del Giudice is nine miles.

And also from the castle of Carunchio to Castiglione Messer Marino[42] and from there to Castel di Jonato is twenty-four miles and from Castel di Jonato to Castel di Sangro is twenty-four miles and likewise from Monteferrante to Castiglione Messer Marino is eighteen miles and from Castiglione Messer Marino to Fallo is twenty-four miles.

The city of Fallo is at the base of the *Qarsilūn* mountain and from Fallo to the town of Liscia is twelve miles and this is at the base of Monte Barello.[43] From Liscia to the castle of Cerro mentioned previously is fifteen miles and it is at the base of the mountain already mentioned. From Monte Melo to the castle of Agnone is eighteen miles and it is at the base of the mountain already mentioned.[44] From Guardiagrele to the castle of Carunchio is eighteen miles and it is at the top of the mountain already mentioned.

From the castle of Carunchio to the castle of Civitaluparella is thirty-three miles and this is a well-populated castle which resembles a city and is located in a gully between Monte Barello and Monte Maiella. And between Civitaluparella and Pennapiedimonte, which is a castle, is twenty-four miles and between Pennapiedimonte and Lama dei Peligni which has already been mentioned is fifteen miles and from Lama dei Peligni to Taranto is six miles.

We return and say that from the castle of Civitaluparella to the castle of Prata is eighteen miles and from Prata to the castle of Pretoro is eighteen miles and this is a pleasant castle at the base of Monte Maiella already mentioned.[45] From Pretoro to the castle of Palena is thirty-six miles and it is also at the base of the aforementioned Monte Maiella. From Palena to *Rocca Qūrālī*[46] is twenty-four miles and it is at the base of the mountain already mentioned and from *Rocca Qūrālī* to the city of *Nūblū* is twenty-one miles then to the castle of Fara San Martino is fifteen miles and it is also at the base of the mountain already mentioned.[47]

From Fara San Martino to the castle of Altino indirectly is twelve miles and this is at the base of the mountain already mentioned. From there to Fara di Pescara is nine miles and from there to Ortona on the coast which is also known as *Aṭrāna* with *ṭa*, is eighteen miles.

**42** Distance missing from all manuscripts.

**43** The location of *Qarsilūn* is uncertain.

**44** Monte Mello is located in the Monti Pizzi range, now part of the Parco Nazionale della Maiella.

**45** Prata is the Torretta di Prata, also known as the Torretta di Casoli.

**46** Location uncertain. Amari and Schiaparelli suggest Roccascalegna. Al-Idrisi, *L'Italia descritta nel "Libro del Re Ruggero"*, 119 and Bresc and Nef Roccamorice, al-Idrisi, *La Première géographie de l'Occident*, 394.

**47** The location of *Nūblū* is uncertain. Bresc and Nef identify this as Casale Nobile, four kilometres north of Altino. Al-Idrisi, *La Première géographie de l'Occident*, 394n2.

We return and say that from Castiglione to the castle of Agnone previously mentioned is fifteen miles and from Agnone to Rocca del Abate is eighteen miles and from there to Civitaluparella is thirty-three miles.

The castle of Civitaluparella is the first place in the lands of the Romans. From Civitaluparella to the castle of Altino is twenty-four miles and Altino is at the base of the mountain already mentioned. From Altino to the castle of *Nūblū* is twelve miles and it is at the base of the mountain already mentioned. From *Nūblū* to Lanciano [*Ankazma*] is twenty-four miles and from there to Ortona, and this is Ortona on the coast as we stated previously, is thirty-six miles.

And from Castiglione to Calcasacco is five days travel and from Campomarino to Guardialfiera is twelve miles and this is a small, civilized city and from there to Castiglione is twenty-four miles.

And from Palmoli to Calcasacco is eighteen miles and from the castle of Calcasacco to the castle of Archiano is twelve miles and likewise from the castle of *Buna Dabūnsī* to the castle of Gissi is eighteen miles and from Gissi to Palmoli is fifteen miles and from Gissi to Carunchio is twelve miles and from the castle of Carunchio to Pennadomo also known as *Tūtlī* is fifteen miles.[48]

And from the city of Ortona to the castle of *Lantishak* is six miles, from there to Pennadomo is fifteen miles and from there to Roccamontepiano is twenty-one miles.[49] And from *Lantishak* to Canosa Sannita [*Fawniṣa*][50] is nine miles and from Canosa Sannita to Pennapiedimonte is fifteen miles.

From Ortona to Rapino is nine miles and from there to Canosa Sannita is nine miles and from Rapino to Roccamontepiano is twelve miles and from there to Pennadomo is nine miles. From the city of Ortona which is on the sea to Bucchianico is twelve miles and from there to Rapino is twelve miles.

From Bucchianico to Chieti [*Tāt*] is twelve miles and Ortona to Chieti is fifteen miles and Chieti is nine miles from the river.[51] And from Chieti to Lanciano is twelve miles and from Lanciano to Bucchianico is nine miles and also from Lanciano to Ortona on the sea is thirty-six miles.

Then we return and say that from the city of Taranto to Matera is one hundred and eighty miles in a north-westerly direction. Matera is a pleasant city with a large territory and is well-developed. From Matera to Bari in an eastward direction is one hundred and eighty miles. And from Matera to the city Gravina is sixty miles in a north-westerly direction. Gravina is a civilized city with a small but fertile territory and from Gravina to Venosa is one hundred and eighty miles.

Venosa is a famous city within the countries of the Lombards and from there to Bari is sixty-five miles eastward. And from Venosa to Andria is fifty-four miles eastward

---

**48** The location of *Buna Dabūnsī* is uncertain.

**49** Bresc and Nef identify *Lantishak* as Lentisco. Al-Idrisi, *La Première géographie de l'Occident*, 395.

**50** Al-Idrisi uses *fa* instead of *qaf* in this instance.

**51** The Pescara River.

and from Andria to Ortona on the coast previously mentioned is forty-five miles eastward and from the city of Venosa also to Irsina [*Munt Fanjūs*] is seventy miles.[52]

And from Irsina to Altogianni is six miles and from Altogianni to Matera is twelve miles.[53] Irsina is a pleasant city with many fertile vineyards and trees. And from Irsina to Grottole is eighteen miles and this is a small, civilized city and from Grottole to Altogianni is six miles and from there to Tricarico which is also called *Itrīkirkū* is eighteen miles.

From Altogianni to Irsina is six miles and from Tricarico to Aliano [*Islān*] is twenty-seven miles and from Aliano to Anglona is twenty-four miles and from Irsina to Tricarico is sixty-three miles in a westerly direction and from Tricarico to the city of Acerenza is seventy-two miles. From the city of Acerenza to the city of Oppido Lucano [*Libūtrāy*] is eighteen miles.[54]

From Acerenza to Potenza is sixty miles and this is an important city with a large, well-populated territory with many vineyards and cultivated fields. From Potenza to Montecalvo Irpino is one hundred and fifty miles in a westerly direction.

And from Potenza to the city of inland Melfi is fifty-four miles in a westerly direction and from inland Melfi to Conza della Campania is one hundred and eight miles and from Conza to Campania is sixty miles. From Campania to Eboli is twenty-seven miles in a westerly direction and from Eboli to the city of Salerno is seventy-two miles.

And also from the city of Campania, which is a large and well-populated castle, to the castle of Balvano [*Balqas*] is seventy-two miles and from the castle of Balvano to the castle of Teggiano [*Diyāna*] is seventy-two miles and from Teggiano to Capaccio and from Capaccio to Salerno is thirty-six miles.[55]

Now we return and mention the road from the city of inland Melfi to the city of Lesina which is on the Adriatic coast. From inland Melfi to Sant'Agata di Puglia in a north-westerly direction is fifty-four miles and from Sant'Agata di Puglia to the castle of Ascoli Satriano is thirty-six miles and it is also called *ʿAzqala* and *ʿAsqala*.

From Ascoli Satriano to Ordona is fifty-four miles and from Ordona to San Lorenzo is fifty-four miles and from there to Foggia is twenty-seven miles. From Foggia to Castelnuovo is sixty-three miles and from Castelnuovo to San Chirico Raparo is thirty-six miles. From there to San Severo is fifty-four miles and from San Severo to the city of Lesina is sixty-three miles.

The city of Lesina is close to the Adriatic Sea. The road from Ordona to Salerno is via the city of Benevento, from Ordona to the city of Adriano is sixteen miles and from Adriano to the city of Apice is eighteen miles and from there to the large city

**52** Irsina was formerly Montepeloso, site of the Battle of Montepeloso (1041) between the forces of the Byzantine Empire against the coordinated forces of Normans and Lombards, the latter of whom were successful in expelling the Byzantines from the region.

**53** Altogianni is also known as Altojanni located near Castel Crotula. Massimo Osanna, Dimitris Roubis, and Francesca Sogliani, "Lo scavo archeologico del sito fortificato medievale di Altojanni (Grottole–MT)," *Siris* 8 (2007): 137–56.

**54** Oppido Lucano was ancient Oppidum and from 1863–1933 Palmira.

**55** Teggiano was ancient Tegianum, then Diano until 1862.

of Benevento is twenty-seven miles and from Benevento to Avellino is seventy-two miles.

From Avellino to Salerno is seventy-two miles and also from Benevento to Cancello is twenty-seven miles and from Cancello to Naples is thirty-six miles and also from Benevento to Montesarchio is fifty-four miles and from Montesarchio to Airola is fifty-four miles. From Airola to Arienzo is forty-two miles and from Arienzo to Cancello is twenty-one miles. From Cancello to Palma Campania is twelve miles and from Palma to Sarno [*Sarna*] is thirty-six miles. From Sarno to Salerno is seventy-two miles.

The road from Ortona to Ancona. From Ortona to the Foro [*Lūqdū*] River is seventy-five miles and from there to the castle of *Anziqa*[56] is twenty miles and from there to the Pescara River is ninety miles. From the Pescara River to *Baṭlān* is ninety miles and from there to Tronto is sixty-nine miles and from Tronto to the city of Numana is one hundred and seventy-four miles and from Numana to the city of Ancona is eighteen miles.[57]

We have now sufficiently mentioned the localities and routes of this part, now we will mention its large running rivers with all vigour and diligence. The first of these rivers is called the Senise River.

And we say that the Senise River comes out from the mountains of Carbone and runs between Calvera and Castronuovo di Sant'Andrea and then flows a short way before joining with the Sinni [*Sakna*] River. It then flows in front of Favale and then proceeds opposite *Shant Bārid Kamīra* and from there to the sea.[58]

The Sinni River comes from the mountain of Sirino and flows until it meets the Senise River and the two unite, it then flows to Favale and *Shant Bārid Kamīra* and then on to the sea. The mountain of Sirino is opposite Viggiano, located between Chiaramonte and Viggiano. Between the mountain just mentioned and Chiaramonte is twelve miles and between the mountain and Viggiano is fifteen miles.

The Agri River comes out from the western side of the mountain of Sirino then flows to Sarconi and passes San Martino d'Agri then proceeds to the castle of Aliano, passing Anglona and on to the castle of Policoro and from there to the sea.

The river of Potenza which is called Basento comes from the mountain near Potenza and then flows adjacent to the city of Tricarico then descends to the eastern side of the town called Grottole [*Akrit*]. The river then further descends to the eastern side of the town called Miglionico and between the river and Miglionico is around four and a half miles. The river then descends to the church of San Teodoro Vecchio then flows to the town called Pomarico Vecchio [*Tartighrīr*] adjacent to the town's right side then descends to the sea.

---

**56** Amari and Schiaparelli suggest *Anziqa* may be Francavilla al Mare. Al-Idrisi, *L'Italia descritta nel "Libro del Re Ruggero"*, 125.

**57** The location of *Baṭlān* is uncertain.

**58** Bresc and Nef suggest *Shant Bārid Kamīra* could be Camarda (now Bernalda). Al-Idrisi, *La Première géographie de l'Occident*, 397.

The Bradano River is also known as the *Abrāṭnū*. The river flows on its own and does not meet with the Basento River. At its origin the river divides into two small rivers which flow between two towns, the first town is called Lagopesole [*Liqbāra*] and the second is called Potenza. These two rivers join again at the rock of Giuliano [*Ruqqa Fanjalān*] and it is from this point that the river is known by the name Bradano. The river then proceeds through cultivated land to the sea. On the banks of this river are plentiful pine trees, the pine is cut and loaded onto the river and sent to the seaside. Pitch and tar can be extracted from the pine and exported to various countries.

The Fortore River comes from the mountain of *Qatrashāl*[59] and descends to *Rība* with *Rība* on the river's right side, and between the river and *Rība* is three miles.[60] The river then passes Castello Mano with Castello Mano on the river's right side and between Castello Mano and the river is a stone's throw. The river proceeds and passes Tufara leaving it on its left side and there are three miles between the river and the town. The river then passes the city of Celenza Valfortore leaving the city on its right and between the river and the town is three miles. The river then passes Macchia and between Macchia and the river is one mile. The river then flows past *Shant Jūwān Mayūr* and between the town and the river is a distance of three miles.[61] The river then passes on the right side of *Lūrant* and between the town and the river is about three miles.[62] The river then passes the right side of Dragonara touching its walls then flows to the right side of Civitate and between Civitate and the river is approximately an arrow's throw. The river then passes the right side of Ripalta and between Ripalta and the river is a stone's throw. Between Ripalta and the sea the river passes close to the eastern side of Lesina which was mentioned previously. Between the mouth of the river and Lesina there is a distance of three miles.

From the mouth of this river in a westerly direction is the small city called Campomarino and between this city and the river is eighteen miles and between Campomarino and the sea is nine miles.

The Neto River stems from the Sila on the right side of Cerenzia and runs eastward. Another river comes from the left side of the area just mentioned to meet with the Neto River in a place known as the Saltworks.[63] And between the meeting place of the two rivers and Cerenzia, also known as *Jirinsiya*, mentioned above, is nine miles. Then the river extends until it passes Santa Severina and between the river and this town is a mile and a half. The river then flows until it passes between Crotone and Strongoli then it empties into the sea.

---

**59** Amari and Schiaparelli suggest *Qatrashāl* could be the feudal seat of Vetestrello (ancient Vescellium, known as Vetrascellum in the *Catalogus baronum*) south of Biccari. Al-Idrisi, *L'Italia descritta nel "Libro del Re Ruggero"*, 127n5.

**60** According to Bresc and Nef *Rība* is Ripa d'Altina near San Bartolomeo in Galdo. Al-Idrisi, *La Première géographie de l'Occident*, 398n1.

**61** *Shant Jūwān Mayūr* is is literally "San Giovanni Maggiore." The location is uncertain.

**62** *Lūrant* is possibly Loreto Aprutino.

**63** Location uncertain.

The Agri River does not run between Sant'Arcangelo and Tursi rather it runs parallel to them and between the river and Tursi is a mile and a half and between the river and Sant'Arcangelo no distance at all. Between Tursi and Montemurro is twelve miles and the river already mentioned runs touching it.[64] The Agri river stems from the mountain of Marsico and between the source of the river and Montemurro is eighteen miles and from there it reaches Tursi and the city of Sant'Arcangelo as we have already stated.

The Sabato River comes out between two rocks on the mountain opposite the northern side of Serino and between Serino and the river is around a mile and half. The river continues on and passes the right side of the castle of Montoro then it moves on to San Barbato and between there and the river is a distance of one and a half miles. The river then runs under the northern side of Atripalda and between Atripalda and the river is a mile and a half. The river then passes under the mountain of Montefredano [*Ibrāṭna*], passing on the right side of the castle of Tufo also called *Ṭuf* and between Tufo and the river is an arrow's throw. The river then flows under Roccabascerana then proceeds under the castle of Ceppaloni and between there and the river is a distance of a mile and half. The river then passes the castle of Montefosco and its territories and they are on the right side of the river. The river already mentioned passes between the territory of Montefosco and Ceppaloni and finishes at the village of Benevento, again located on the right side of the river.

The Calore River's source is the mountain called Monte Terminio, the river then passes Taurasi and between Taurasi and the river there are mountains and valleys. The river passes the estate of Sant'Anna [*Raḥl Qanṭāna*] and between there and the river is a mile and a half. The river then flows under the northern side of Montefosco at a distance of six miles.

This river proceeds between the castle of Mirabella Eclano[65] [*Iqtaranda*] and Montefosco which is located to the right side of the river and between the two places and the river is a distance of a mile and half. The river then passes the right side of Apice and between there and the river is the distance of a stone's throw. The river passes also under Padula which is located to the right of the river and between the river and the mountain[66] is a mile and a half and between the mountain and Padula is half a mile.[67] The river then passes under the Valentino bridge and between the bridge and Benevento is three miles. Finally, the river meets the Sabato river at San Vitale.

The Lao River comes out and descends opposite Mercuri Tedesco to the region that is opposite Scalea and then to the sea. The Rivello River comes out of a hill in the vicinity of Castrocucco and Maratea. From there the river descends to the sea. Between Scalea and the river mentioned is six miles and from the river to Maratea is one mile.

---

**64** Montemurro? This is unclear.

**65** Mirabella Eclano was previously Aquaputida.

**66** Al-Idrisi doesn't specify which mountain.

**67** The Valentino is a Roman bridge.

The Bussento River descends from opposite Scanza [*Shant Ṣītrī*] to Policastro and from there to the sea. While the Mingardo River descends from *Qastilnas* mountain to opposite Camerota then descends to opposite Mingardo to the sea.[68]

The Alento [*Abrūqa*] river descends from Monteforte Cilento and from there proceeds to opposite Corbella then proceeds to opposite Torchiara and from there to Castellammare della Velia and then the sea.

The Tanagro [*Adyāna*] River comes out at Montesano then descends close Padula then it proceeds between Diano and Sala Consilina then comes out at Polla, proceeding then opposite Auletta and close to Cintursi then joins with the Sele River.

The Caselle River, also known with 'ha' as *Qazluh*, descends from the *Akrān* monastery then proceeds to Caselle and empties into the Policastro River.[69] While the Rocca Gloriosa River descends also from the mountains to the front of Rofrano then proceeds to Rocca Gloriosa then to Camerota and finally empties into the sea.

We have now completed the Third Part of the Fifth Clime, thanks be to Allah, and we will now proceed to the Fourth Part, Allah willing.

---

**68** The location of *Qastilnas* is uncertain.

**69** The location of *Akrān* is uncertain.

# SELECT BIBLIOGRAPHY

## Manuscripts of the *Nuzhat al-Mushtaq*

This listing of manuscripts is taken from Appendix 7.1 of Ahmad's "Cartography of al-Sharif al- Idrīsī" (cited in full below). Details about the date and location of creation, state of preservation, and existence of maps within these manuscripts can be found there as well.

Cairo, Dār al-Kutub, Jugrāfiyā 150 (Egyptian Library, Gezira 150; Kat. Vs. 167).
Istanbul, Köprülü Kütüphanesi, MS 955 (Ğuġrāfiyā 702).
Istanbul, Süleymaniye Kütüphanesi, Ayasofya 3502 (Hagia Sofya 3502 ; Ğuġrāfiyā 705)
Leningrad, M.E. Saltykov-Shchedrin State Public Library, MS Ar. N.S. 176. (Saint Petersburg, Cod. Arab. 4, 1, 64).
London, India Office, Loth 722 (MS Ar. 617).
Oxford, Bodleian Library, MS Greaves 42 (MS Greaves 3847-42; Uri 884).
Oxford, Bodleian Library, MS Pococke 375 (Uri 887).
Paris, Bibliothèque nationale de France , MS Arabe 2221 (Suppl. MS Arabe 892).
Paris, Bibliothèque nationale de France , MS Arabe 2222 (Suppl. MS Arabe 893).
Sofia, Cyril and Methodius National Library, MS Or. 3198 (MS Or. 3180).

## Critical Edition of the *Nuzhat al-Mushtaq* used for Translation

al-Idrisi. *Opus geographicum, sive "Liber ad eorum delectationem qui terras peragrare studeant"*. Edited by Enrico Cerulli. 9 vols. Naples: Brill, 1970–84.

## Partial Translations of the *Nuzhat al-Mushtaq*

al-Idrisi. *Description de l'Afrique et de l'Espagne Par Edrisi*. Translated by Reinhart Dozy and Michael Jan Goeje. Leiden: Brill, 1866.
al-Idrisi. *Géographie d'Édrisi*. Translated by Amédée Jaubert. 2 vols. Recueil de Voyages et de Mémoires 5. Paris: Libraire de la Societé, 1836.
al-Idrisi. *L'Italia descritta nel "Libro del Re Ruggero."* Translated by Michele Amari and Celestino Schiaparelli. Rome: Tipi del Salviucci, 1883.
al-Idrisi. *Il Libro di Ruggero*. Translated by Umberto Rizzitano. Palermo: Flaccovio, 1966.
al-Idrisi. *La Première géographie de l'Occident*. Translated by Henri Bresc and Annliese Nef. Paris: GF Flammarion, 1999.

## Dictionaries Consulted

Hava, J. G. *Al-Fara'id Arabic–English Dictionary*. Beirut: Dar al-Mashriq, 1970.
Ibn Manzur. *Lisan al-ʿArab*. 3rd ed. 20 vols. Dar Sadir, 2000.
Lane, Edward. *Arabic–English Lexicon*. 8 vols. London: Williams & Norgate, 1863.
Wehr, Hans. *Arabic–English Dictionary: The Hans Wehr Dictionary of Modern Written Arabic*. Edited by J. M. Cowan. 4th ed. Ithaca: Spoken Language Services, 1994.

## Selected Other Sources


Abulafia, David. "Ancona, Byzantium and the Adriatic, 1155–1173." *Papers of the British School at Rome* 52 (1984): 195–216.

——. "Local Trade Networks in Medieval Sicily: The Evidence of Idrisi." In *Shipping, Trade and Crusade in the Medieval Mediterranean: Studies in Honour of John Pryor*, edited by Ruthy Gertwagen and Elizabeth Jeffreys, 157–66. Taylor & Francis, 2012.

——. *The Two Italies: Economic Relations between the Norman Kingdom of Sicily and the Northern Communes*. New York: Cambridge University Press, 1977.

Adams, J. M. *The Regional Diversification of Latin 200 BC–AD 600*. New York: Cambridge University Press, 2007.

Ahmad, S. Maqbul. "Cartography of Al-Sharīf al-Idrīsī." In *Cartography in the Traditional Islamic and South Asian Societies*, edited by J.B. Harley and David Woodward, 156–74. The History of Cartography 2.1. Chicago: University of Chicago Press, 1992.

al-Bakri. *Al-Masalik wa-l-Mamalik*. 2 vols. Beirut: Dar al-Kotob al-Ilmiyah, 2003.

——. *Description de l'Afrique septentrionale par el-Bekri*. Translated by William McGuckin de Slane. Paris: Imprimerie Impériale, 1859.

Alexander of Telese. *Alexandri Telesini Abbatis Ystoria Rogerii Regis Sicilie Calabrie atque Apulie*. Edited by Ludovica de Nava. Fonti per la Storia d'Italia 112. Rome: Instituto Palazzo Borromini, 1991.

al-Hamawi, Yaqut. *Muʿjam Al-Buldan*. 6 vols. Beirut: Dar Sadir, 1995.

al-Maqdisi. *Aḥsan al-taqāsīm fī maʿrifat al-aqālīm*. Cairo: Maktaba Madbouly, 1991.

——. *The Best Divisions for Knowledge of the Regions: Ahsan al-Taqasim Fi Marifat al-Aqalim*. Translated by Basil Anthony Collins. Reading: Garnet, 2000.

al-Maqrizi. *Kitab al-Muqaffa al-Kabir*. Edited by Muhammad al-Yaʿlawi. 8 vols. Beirut: Dar al-Gharb al-Islami, 1991.

al-Safadi. *Kitab al-Wafi bi-l-Wafayat*. Edited by Ahmad al-Arnåut and Turki Mustafa. 29 vols. Beirut: Dar Ihyå al-Turath al-ʿArabi, 2000.

al-ʿUmari. *Al-Taʿrīf Bi-al-Mustalah al-Sharif*. Beirut: Dar al-Kotob al-Ilmiyah, 1988.

Amara, Allaoua, and Annliese Nef. "Al-Idrisi et les Hammudides de Sicile: Nouvelles donnees biographiques sur l'auteur du livre de Roger." *Arabica* 48 (2001): 121–27.

Amari, Michele. *Biblioteca Arabo-Sicula*. Leipzig: F.A. Brockhaus, 1857.

——. *Biblioteca Arabo-Sicula, Versione Italiana*. 3 vols. Torino: Ermanno Loescher, 1880.

——. *Storia dei Musulmani di Sicilia*. 3 vols. Florence: Felice le Monnier, 1858.

Anderson, Glaire D., Corisande Fenwick, and Mariam Rosser-Owen, eds. *The Aghlabids and Their Neighbors: Art and Material Culture in Ninth-Century North Africa*. Leiden: Brill, 2017.

Antrim, Zayde. *Routes and Realms: The Power of Place in the Early Islamic World*. New York: Oxford University Press, 2012.

Aristippus, Henricus. *Phaedo interprete Henrico Aristippo*. Edited by Laurentius Minio-Paluello. Plato Latinus. London: Warburg Institute, 1950.

Baadj, Amar. *Saladin, the Almohads and the Banū Ghāniya: The Contest for North Africa (12th and 13th Centuries)*. Leiden: Brill, 2015.

Babelon, Ernest, René Cagnat, and Salomon Reinach. *Atlas archéologique de la Tunisie*. Paris: Ministère de la Guerre, 1893.

Bagnera, Alessandra. "The Urban Evolution of Islamic Palermo (9th–Mid-11th Century)." In *A Companion to Medieval Palermo*, edited by Annliese Nef, 61–88. Leiden: Brill, 2013.

Balard, Michel. "Notes sur le commerce entre l'Italie et l'Égypte sous les Fatimides." In *L'Égypte fatimide: son art et son histoire*, 627–33. Paris: Presses de l'Université de Paris-Sorbonne, 1999.

Bates, David, and Pierre Baudin, eds. *911–2011: Penser les mondes normands médiévaux*. Caen: Presses universitaires de Caen, 2016.

Beeston, A. F. L. "Idrisi's Account of the British Isles." *Bulletin of the School of Oriental and African Studies* 13 (1950): 265–80.

Birk, Joshua. *Norman Kings of Sicily and the Rise of the Anti-Islamic Critique: Baptized Sultans*. New York: Palgrave Macmillan, 2016.

Bramoullé, David. *Les Fatimides et La Mer (909–1171)*. Leiden: Brill, 2019.

Bresc, Henri. "La Sicile et l'espace Libyen au moyen-âge." *Africa: Rivista trimestrale di studi e documentazione dell'Istituto italiano per l'Africa e l'Oriente* 63 (2008): 187–208.

Brett, Michael. *The Fatimid Empire*. Edinburgh: Edinburgh University Press, 2017.

——. "Ifriqiya as a Market for Saharan Trade from the Tenth to the Twelfth Century AD." *The Journal of African History* 10 (1969): 347–64.

——. *The Rise of the Fatimids: The World of the Mediterranean and the Middle East in the Fourth Century of the Hijra, Tenth Century CE*. Leiden: Brill, 2001.

Britt, Karen. "Roger II of Sicily: Rex, Basileus, and Khalif? Identity, Politics, and Propaganda in the Cappella Palatina." *Mediterranean Studies* 16 (2007): 21–45.

Cambuzat, Paul-Louis. *L'évolution des cites du Tell en Ifrîkiya du VII.° au XI.° siécle*. 2 vols. Argel: Office des Publications Universitaires, 1986.

Constable, Olivia Remie. *Housing the Stranger in the Mediterranean World: Lodging, Trade, and Travel in Late Antiquity and the Middle Ages*. New York: Cambridge University Press, 2003.

Curta, Florin. *Southeastern Europe in the Middle Ages, 500–1250*. Cambridge: Cambridge University Press, 2006.

Dalli, Charles. "Bridging Europe and Africa: Norman Sicily's Other Kingdom." In *Bridging the Gaps: Sources, Methodology and Approaches to Religion in History*, edited by Joaquim Carvalho, 77–93. Pisa: Pisa University Press, 2008.

Davis-Secord, Sarah. "Sicily and the Medieval Mediterranean: Communication Networks and Inter-Regional Exchange." PhD diss., Notre Dame, 2007.

——. *Where Three Worlds Met: Sicily in the Early Medieval Mediterranean*. Ithaca: Cornell University Press, 2017.

Di Branco, Marco, and Kordula Wolf. "Hindered Passages: The Failed Muslim Conquest Of Southern Italy." *Journal of Transcultural Medieval Studies* 1 (2014): 51–74.

Dolezalek, Isabelle. *Arabic Script on Christian Kings: Textile Inscriptions on Royal Garments from Norman Sicily*. Berlin: De Gruyter, 2017.

Drell, Joanna, and Paul Oldfield, eds. *Rethinking Norman Italy: Studies in Honour of Graham A. Loud*. Manchester: Manchester University Press, 2021.

Ducène, Jean-Charles. *L'Afrique dans le Uns al-Muhağ Wa-Rawḍ al-Furağ d'al-Idrīsī*. Leuven: Peeters, 2010.

——. "Al-Idrīsī, Abū ʿAbdallāh." In *Encyclopaedia of Islam*. Leiden: Brill, 2018. http://dx.doi.org/10.1163/1573-3912_ei3_COM_32372.

——. "Les œuvres géographiques d'al-Idrīsī et leur diffusion." *Journal Asiatique* 305 (2017): 33–41.

——. "Routes in Southern Italy in the Geographical Works of Al-Idrīsī." In *Journeying along Medieval Routes in Europe and the Middle East*, edited by Alison Gascoigne, Leonie Hicks, and Marianne O'Doherty, 143–66. Turnhout: Brepols, 2016.

Dyson, Stephen L., and Robert J. Rowland, Jr. *Archaeology and History in Sardinia from the Stone Age to the Middle Ages: Shepherds, Sailors & Conquerors*. Philadelphia: University of Pennsylvania Press, 2007.

Falasca, Vincenzo. *Il castello Sanseverino di Grumento Nova*. Grumento Nova: Patrocinio Amministrazione Communale, 2010.

Falcandus, Hugo. *La Historia o Liber de Regno Sicilie e la Epistola ad Petrum Panormitane Ecclesie Thesaurarium*. Edited by G. B. Siragusa. Fonti per la Storia d'Italia. Rome: Tipografi del Senato, 1897.

——. *The History of the Tyrants of Sicily by "Hugo Falcandus," 1154–69*. Translated by Graham Loud and Thomas Wiedemann. New York: Manchester University Press, 1998.

Fauvelle, François-Xavier. *The Golden Rhinoceros: Histories of the African Middle Ages*. Translated by Troy Tice. Princeton: Princeton University Press, 2018.

Feller, Laurent. "The Northern Frontier of Norman Italy, 1060–1140." In *The Society of Norman Italy*, edited by Graham Loud and Alex Metcalfe, 47–74. Leiden: Brill, 2002.

Fine, John Van Antwerp. *The Early Medieval Balkans: A Critical Survey from the Sixth to the Late Twelfth Century*. Ann Arbor: University of Michigan Press, 1991.

Garali, Amor Ben, Mohamed Ouakad, and Moncef Gueddari. "Geochemistry and Ionic Interaction in the Bizerte Lagoon Waters (Northern Tunisia)." *Journal of Oceanography and Marine Science* 2, no. 1 (January 2011): 1–9.

Geoffrey Malaterra. *The Deeds of Count Roger of Calabria and Sicily and of His Brother Duke Robert Guiscard*. Translated by Kenneth Baxter Wolf. Ann Arbor: University of Michigan Press, 2005.

——. *De rebus gestis Rogerii Calabriae et Siciliae comitis et Roberti Guiscardi ducis fratris eius*. Edited by L.A. Muratori and Ernersto Pontieri. Rerum Italicarum Scriptores 5. Bologna: Nicola Zanichelli, 1949.

Gertwagen, Ruthy. "Geniza Letters: Maritime Difficulties along the Alexandria-Palermo Route." In *Communication in the Jewish Diaspora: The Pre-Modern World*, edited by Sophia Menache, 73–90. Leiden: Brill, 1996.

Geus, Klaus. "Oikoumene/Orbis Terrarum." In *Oxford Classical Dictionary*, 2016. https://doi.org/10.1093/acrefore/9780199381135.013.8008.

Goitein, S. D. *Letters of Medieval Jewish Traders*. Princeton: Princeton University Press, 1973.

Goldenberg, David M. *The Curse of Ham: Race and Slavery in Early Judaism, Christianity, and Islam*. Princeton: Princeton University Press, 2009.

Granara, William. *Ibn Hamdis the Sicilian: Eulogist for a Falling Homeland*. London: Oneworld Publications, 2021.

Green, Judith A. *The Normans: Power, Conquest & Culture in 11th-Century Europe*. New Haven: Yale University Press, 2022.

Grierson, Philip, and Lucia Travaini. *Italy III: South Italy, Sicily, Sardinia*. Medieval European Coinage 14. New York: Cambridge University Press, 1998.

Gsell, Stéphane. *Atlas archéologique de l'Algérie*. Algiers: A. Jourdan, 1911.

Guérin, Sarah M. "Forgotten Routes? Italy, Ifrīqiya and the Trans-Saharan Ivory Trade." *Al-Masāq: Islam and the Medieval Mediterranean* 25 (2013): 70–91.

Halm, Heinz. *The Empire of the Mahdi: The Rise of the Fatimids*. Leiden: Brill, 2021.

Hannoum, Abdelmajid. "Translation and the Colonial Imaginary: Ibn Khaldūn Orientalist." *History and Theory* 42, no. 1 (February 2003): 61–81.

Hayes, Dawn Marie. "The Cult of St Nicholas of Myra in Norman Bari, ca. 1071–ca. 1111." *The Journal of Ecclesiastical History* 67 (2016): 492–512.

——. *Roger II of Sicily: Family, Faith, and Empire in the Medieval Mediterranean World*. Turnhout: Brepols, 2020.

Hodges, Richard, William Bowden, and Kosta Lako. *Byzantine Butrint: Excavations and Surveys 1994–99*. Oxford: Oxbow Books, 2020.

Hoffman, Valerie. *The Essentials of Ibadi Islam*. Syracuse: Syracuse University Press, 2012.

Horden, Peregrine, and Nicholas Purcell. *The Corrupting Sea: A Study of Mediterranean History*. New York: Wiley-Blackwell, 2000.

Houben, Hubert. *Roger II of Sicily: A Ruler between East and West*. Translated by Graham Loud and Diane Milburn. New York: Cambridge University Press, 2002.

Hoyos, Dexter. *Carthage: A Biography*. New York: Taylor & Francis, 2020.

Hughes, R. H., J. S. Hughes, and G. M. Bernacsek. *A Directory of African Wetlands*. Cambridge: IUCN, UNEP, and WCMC, 1992.

Ibn al-Athir. *The Chronicle of Ibn Al-Athir for the Crusading Period from al-Kāmil Fī'l-Ta'rīkh*. Translated by D. S. Richards. Vol. 1. Aldershot: Ashgate, 2006.

——. *Al-Kamil fi al-Tarikh*. Beirut: Dar Sadir, 2009.

Ibn Jubayr. *Rihla Ibn Jubayr*. Beirut: Dar Sadir, 1964.

——. *The Travels of Ibn Jubayr: A Medieval Journey from Cordoba to Jerusalem*. Edited by Robert Irwin. Translated by Ronald Broadhurst. New York: I. B. Tauris, 2019.

Ibrahim, Sufyan Yasen. "Exploration of the Geographical Location of Waq Waq Islands In The Light of Islamic Arab Geographical Sources." *Alustath Journal for Human and Social Science* 2, no. 225 (2018): 51–64.

Idris, H. R. *La Berbérie orientale sous les Zīrīdes: Xe–XIIe siècles*. 2 vols. Limoges: A. Bontemps, 1962.

Irwin, Robert. *Ibn Khaldun: An Intellectual Biography*. Princeton: Princeton University Press, 2018.

Jacoby, David. "Silk Crosses the Mediterranean." In *Le Vie del Mediterraneo: Idee, Uomini, Oggetti (Secoli XI–XVI)*, edited by Gabriella Airaldi, 55–79. Genoa: ECIG, 1997.

Johns, Jeremy. *Arabic Administration in Norman Sicily: The Royal Dīwān*. New York: Cambridge University Press, 2002.

——. "The Norman Kings of Sicily and the Fatimid Caliphate." *Anglo-Norman Studies* 15 (1993): 133–59.

Kaegi, Walter E. *Muslim Expansion and Byzantine Collapse in North Africa*. New York: Cambridge University Press, 2010.

Kahlaoui, Tarek. *Creating the Mediterranean: Maps and the Islamic Imagination*. Leiden: Brill, 2017.

——. "Ibadi Jerba: Surviving the Early Ḥafṣids (13–14th Centuries)." *The Journal of North African Studies* 26 (2021): 631–41.

Kennedy, Edward S. "Mathematical Geography." In *The Encyclopedia of the History of Arabic Science*, edited by Roshdi Rashed, 1:187–88. New York: Routledge, 1996.

Kreutz, Barbara. *Before the Normans: Southern Italy in the Ninth and Tenth Centuries*. Philadelphia: University of Pennsylvania Press, 1991.

Krueger, Hilmar. "The Italian Cities and the Arabs before 1095." In *A History of the Crusades*, edited by Marshall W. Baldwin and Kenneth Meyer Setton, 1:40–54. Madison: University of Wisconsin Press, 1969.

Lev, Yaacov. "A Mediterranean Encounter: The Fatimids and Europe, Tenth to Twelfth Centuries." In *Shipping, Trade and Crusade in the Medieval Mediterranean: Studies in Honour of John Pryor*, edited by Ruthy Gertwagen and Elizabeth Jeffreys, 131–56. New York: Taylor & Francis, 2012.

Levtzion, Nehemia, and J. F. P. Hopkins, eds. *Corpus of Early Arabic Sources for West African History*. Translated by J. F. P. Hopkins. Princeton: Markus Wiener Publishers, 2000.

Lopez, Robert S., and Irving W. Raymond. *Medieval Trade in the Mediterranean World*. New York: Columbia University Press, 1955.

Loud, Graham. *The Age of Robert Guiscard: Southern Italy and the Norman Conquest*. New York: Longman, 2000.

——. *Roger II and the Creation of the Kingdom of Sicily*. New York: Manchester University Press, 2012.

Love Jr., Paul M. *Ibadi Muslims of North Africa: Manuscripts, Mobilization, and the Making of a Written Tradition*. Cambridge: Cambridge University Press, 2018.

Mallette, Karla. *The Kingdom of Sicily, 1100–1250: A Literary History*. Philadelphia: University of Pennsylvania Press, 2005.

Maurici, Ferdinando. *Castelli medievali in Sicilia: Dai bizantini ai normanni*. Palermo: Sellerio editore, 1992.

Mazuz, Hagga. "Tracing Possible Jewish Influence on a Common Islamic Commentary on Deuteronomy 33:2." *The Journal of Jewish Studies* 67 (2016): 291–304.

McDougall, E. Ann. "The Sahara Reconsidered: Pastoralism, Politics and Salt from the Ninth through the Twelfth Centuries." *African Economic History* 12 (1983): 263–86.

McQueen, William B. "Relations between the Normans and Byzantium 1071–1112." *Byzantion* 56 (1986): 427–76.

Metcalfe, Alex. *Muslims and Christians in Norman Sicily: Arabic Speakers and the End of Islam*. New York: Routledge Curzon, 2003.

——. *The Muslims of Medieval Italy*. Edinburgh: Edinburgh University Press, 2009.

Metcalfe, Alex, Hervin Fernández-Aceves, and Marco Muresu. *The Making of Medieval Sardinia*. Leiden: Brill, 2021.

Mones, Hussain. "Description of Egypt by Idrisi." *Studi Magrebini* 16 (1984): 1–53.

Nef, Annliese. *Conquérir et gouverner la Sicile islamique aux XIe et XIIe siècles*. Rome: École française de Rome, 2011.

——. "Al-Idrīsī : un complément d'enquête biographique." In *Géographes et voyageurs au Moyen Âge*, edited by Henri Bresc and Emmanuelle Tixier du Mesnil, 53–66. Nanterre: Presses universitaires de Paris Nanterre, 2010.

Norie, J.W. *New Sailing Directions for the Mediterranean Sea*. 2nd ed. London: Charles Wilson, 1841.

Noyé, Ghislaine. "Byzantine Calabria." In *A Companion to Byzantine Italy*, edited by Salvatore Cosentino, 434–52. Leiden: Brill, 2021.

Oldfield, Paul. "Urban Government in South Italy ca. 1085–ca. 1127." *The English Historical Review* 122, no. 497 (2007): 579–608.

Passon, Jacqueline, Klaus Braun, Said Hamid, Salih al-Mahdi Khalifa, Najmiya as-Sadeq at-Tellisi, Mansour El Nayedh, and Joschua Metzger. "Places of Trade, Communication and Religion—Important Oases." In *Across the Sahara: Tracks, Trade and Cross-Cultural Exchange in Libya*, edited by Klaus Braun and Jacqueline Passon, 165–247. New York: Springer Open, 2020.

Peacock, Andrew. "Suakin: A Northeast African Port in the Ottoman Empire." *Northeast African Studies* 12, no. 1 (2012): 29–50.

Pryor, John. "Winds, Waves, and Rocks: Routes and the Perils Along Them." In *Maritime Aspects of Migration*, edited by Klaus Friedland, 71–85. Köln: Böhlau, 1989.

Pryor, John, and Elizabeth Jeffreys. *The Age of the Dromon: The Byzantine Navy ca. 500–1204*. Leiden: Brill, 2006.

Ptolemy. *Ptolemy's Geography: An Annotated Translation of the Theoretical Chapters*. Translated by Alexander Jones and J. Lennart Berggren. Princeton: Princeton University Press, 2020.

Romuald of Salerno. *Romualdi Salernitani Chronicon*. Edited by Carlo Garufi. Rerum Italicarum Scriptores 7.1. Citti di Castello: S. Lapi, 1935.

Rouighi, Ramzi. *The Making of a Mediterranean Emirate: Ifriqiya and Its Andalusis, 1200–1400*. Philadelphia: University of Pennsylvania Press, 2011.

Santagati, Luigi. *La Sicilia di Al-Idrisi ne "Il libro di Ruggero"*. Palermo: Sciascia, 2010.

Sezgin, Fuat. *Mathematical Geography and Cartography in Islam and Their Continuation in the Occident*, 3 vols Translated by Guy Moore and Geoff Sammon. Frankfurt am Main: Institute for the History of Arabic-Islamic Science, 2000–2007.

Simonsohn, Shlomo. *Between Scylla and Charybdis: The Jews in Sicily*. Leiden: Brill, 2011.

Smith, David Romney. "Calamity and Transition: Re-Imagining Italian Trade in the Eleventh-Century Mediterranean." *Past & Present* 228, no. 1 (2015): 15–56.

Stanton, Charles. *Norman Naval Operations in the Mediterranean*. Woodbridge: The Boydell Press, 2011.

Stephenson, Paul. *Byzantium's Balkan Frontier: A Political Study of the Northern Balkans*. New York: Cambridge University Press, 2000.

Stevens, Susan T. "Carthage in Transition: From Late Byzantine City to Medieval Villages." In *North Africa under Byzantium and Early Islam*, edited by Susan T. Stevens and Jonathan P. Conant, 89–104. Washington, DC: Dumbarton Oaks, 2016.

Talbi, Mohamed. *L'émirat aghlabide (184/860–296/909)*. Paris: Adrien-Maisonneuve, 1966.

Tolmacheva, Marina A. "Geography." In *Medieval Islamic Civilization: An Encyclopedia*, 1:284–88. New York: Routledge, 2006.

Udovitch, Abraham. "Time, the Sea and Society: Duration of Commercial Voyages on the Southern Shores of the Mediterranean during the High Middle Ages." In *La Navigazione Mediterranea Nell'alto Medioevo*, 503–46. Spoleto, 1978.

Valérian, Dominique. "La permanence du christianisme au Maghreb: l'apport problématique des sources latines." In *Islamisation et arabisation de l'Occident musulman (VIIe–XIIe siècle)*, edited by Dominique Valérian, 131–49. Paris: Éditions de la Sorbonne, 2011.

Wilkinson, John Craven. "A Sketch of the Historical Geography of the Trucial Oman down to the Beginning of the Sixteenth Century." *The Geographical Journal* 130 (1964): 337–49.

Wink, André. *Al-Hind: The Making of the Indo-Islamic World*. 2 vols. Leiden: Brill, 1991.

Yarbrough, Luke B. *Friends of the Emir: Non-Muslim State Officials in Premodern Islamic Thought*. Cambridge: Cambridge University Press, 2019.

# INDEX OF PLACENAMES

THIS INDEX CONTAINS all of the placenames mentioned in our translation. We have chosen to organize this index alphabetically based on our English translation. We have also opted not to include identifiers like "fort of" or "city of" in this alphabetization scheme, since it would cluster entries around a few identifiers, but we have included them as parentheticals. This leads to some peculiar entries. For example, the "castle of the Almond Trees" can instead be found under "Almond Trees (Castle)" and "the fort of the Port of the River" can be found under "Port of the River (Fort)." Next to our translation is the original Arabic text of al-Idrisi and a transliteration according to the standards of the International Journal of Middle East Studies. The locations for these places (when known) is discussed in footnotes when they first occur in our translation. If a particular placename does not have a settlement type after it (such as Estate/City etc.), it means al-Idrisi has not provided this information in his description. We have not included the settlements mentioned in the Preface but have instead focused on the Norman territories in the South which is the subject of this edition.

| Our Translation | Arabic Name | IJMES Transliteration |
|---|---|---|
| Abu Khalifa (Stronghold) | ابي خليفة | Abī Khalīfa |
| Abu Marzuq (Fort) | ابي مرزوق | Abī Marzūq |
| Acerenza (City) | جرسنة | Jarsna |
| Aci Castello (Town) | لياج | Liyāj |
| Acquedolci (Port) | علقمارة | ʿAlqamāra |
| Acri (Town and River) | أكري/أكراك | Akrī/Akrāk |
| Adaru (Mountain) | ادارو | Adārū |
| Adrano (Estate) | أذرنو | Adhranū |
| Adriano (City) | أرنانة | Arnāna |
| Adriatic | خليج البنادقيين | Khalīj al-Banādiqiyīn |
| ʿAfsalat (Fort) | عفسلات | ʿAfsalāt |
| Agira | شنت فيلب | Shant Fīlib |
| Agnone (Castle) | أنقلون | Anqalūn |
| Agnone (Port) | الركن | al-Rakn |
| Agrigento (City) | كركنت | Karkant |
| Agropoli (Gulf) | غروبلي | Gharūblī |
| Ahara (Village) | احرى | Aḥara |
| Ahnas (River) | احناس | Aḥnās |

| Our Translation | Arabic Name | IJMES Transliteration |
|---|---|---|
| Aidone | أيذوني | Aydhūnī |
| Ain Arko (Village) | ازكو | Azkū |
| Aïn Djeloula (City) | جلولة | Jalūla |
| Ain Djoukar (Spring) | شوقار | Shūqār |
| Airola | أبرولة | Abrūla |
| Akrān (Monastery) | أكران | Akrān |
| Akranta | أكرنتة | Akranta |
| Amantea | منتية | Mantiyya |
| al-Anbāṣī (Port) | الأنباصي | al-Anbāṣī |
| Asinara (Island) | أم الحمار | Um al-Ḥimār |
| al-Asnam | الاصنام | al-Aṣnām |
| al-Azqaq (Bay) | الازقاق | al-Azqāq |
| al-Bandari (Fort) | البنداري | al-Bandārī |
| al-Baqla (Cape) | البقلة | al-Baqla |
| al-Bardwan (Village) | البردوان | al-Bardawan |
| al-Daraq (Fort) | الدرق | al-Daraq |
| al-Fawara (Village) | الفوارة | al-Fawara |
| al-Hassan ibn al-Nuʿman al-Ghassani (Forts) | الحسان بن النعمان الغساني | al-Ḥassān ibn al-Nuʿmān al-Ghassānī |
| al-Ifriqi (Fort) | الإفريقي | al-Ifrīqī |
| al-Majtana | المجتنى | al-Majtana |
| al-Mansaf | المنصف | al-Manṣaf |
| al-Māṣa | الماصة | al-Māṣa |
| al-Masin (Cape) | المسن | al-Masin |
| al-Monastir (Forts) | المنستير | al-Monastīr |
| al-Muʿallaqa (Land) | المعلقة | al-Muʿallaqa |
| al-Nahrin (Village) | النهرين | al-Nahrīn |
| al-Qarbadi (Caves) | القربدي | al-Qarbadi |
| al-Qāṭa | القاطة | al-Qāṭa |
| al-Qitana (Castle) | القيطنة | al-Qīṭana |
| al-Rakhima (Cape) | الرخيمة | al-Rakhima |
| al-Safā (Well) | الصفاء | al-Ṣafā |

| Our Translation | Arabic Name | IJMES Transliteration |
|---|---|---|
| al-Ṭarfāniyya (Island) | الطرفانية | al-Ṭarfāniyya |
| al-Tarmidh (Spring) | الطرميذ | al-Ṭarmīdh |
| al-Tifaf (Castle) | التفاف | al-Tifāf |
| al-Zakūjī (Port) | وادي الزكوجي | Wādī al-Zakūjī |
| al-Zarat (Forts) | الزارات | al-Zārāt |
| al-ʿAbbas (Wells) | العباس | al-ʿAbbās |
| Alaâ (Village) | الجهنيين | al-Jahaniyyin |
| Alana (River) | الأنة | Alāna |
| Alcamo (Estate) | علقمة | ʿAlqama |
| Alento | أبروقة | Abrūqa |
| Alessi (River) | اسجلاسة | Isjalāsa |
| Alfedena | ألفدينة | Alfidīna |
| Aliano | اسلان / أليان | Islān / Alyān |
| Alicudi (Island) | انكوذة | Ankūdha |
| Almond (Fort) | اللوزة | al-Lawza |
| Almond Trees (Castle) | الجوزات | al-Jūzāt |
| Almoravids (Fort) | المرابطين | al-Murābiṭīn |
| Altino (Castle) | طن | Ṭin |
| Altogianni | ألت جوان | Alt Jūwān |
| Amalfi (City) | ملف | Malf |
| Amantea (City) | المنتية | Almantiyya |
| Amato (River) | قزليت | Qazlīt |
| Ampurias | وانبوريش | Awnbūrīsh |
| Amrud | امرود | Amrūd |
| Anchorage of ʿAbād (Port) | ترسة عباد | Tarsat ʿAbād |
| Ancinale (River) | بحلاية | Baḥlāyya |
| Ancona (City) | أنكونة / أنقونة | Ankūna / Anqūna |
| Andria (City) | أندرة | Andra |
| Anglona (Castle) | أكلون / أنكلون | Aklūn/Anklūn |
| Annaba (City) | بونة | Būna |
| Antijan (Land) | انتيجان | Antījān |
| Anzio (Port) | أنجة | Anja |

| Our Translation | Arabic Name | IJMES Transliteration |
|---|---|---|
| Anziqa (Castle) | انزقة | Anziqa |
| Apice (City) | أبيج / أبج | Abīj / Abj |
| Apulia (Region) | بولية | Būliyya |
| Aqrabīl | اقربيل | Aqrabīl |
| Aquileia (Region) | ايكلاية | Iklayyia |
| Arienzo (City) | أرجنت | Arjint |
| Armento | أرمنت | Armint |
| Arnaṣ | أرنص | Arnaṣ |
| Arns | أرنس | Arns |
| Asa (River) | سيلسة | Sīlsa |
| Ascoli Satriano (Castle) | عزقلة / عسقلة / أصقلة | ʿAzqala / ʿAsqala / Aṣqala |
| Ashluna (City) | اشلونة | Ashlūna |
| Aspra (Village) | الصبر | al-Ṣabr |
| Astura | أسطونة | Asṭūna |
| Atina | أتينة | Atīna |
| Atripalda | افرنتيو | Ifrintyū |
| Auletta | أولات | Awlāt |
| Aurès (Mountain) | اوراس | Aurās |
| Avellino (City) | أبلينه | Ablīna |
| Aversa | أجرسة | Ajarsa |
| Awjila | اوجلة | Awjila |
| Bābra | بابرة | Bābra |
| Bades (City) | بادس | Bādes |
| Baghai (City) | باغاي | Bāghay |
| Bakar (City) | بقري | Baqrī |
| Bakery (Cape) | المخبز | al-Makhbaz |
| Bakruwa (Port) | باكروا | Bākruwa |
| Balvano (Castle) | بلقس | Balqas |
| Banaqa (Fort) | بنقة | Banaqa |
| Banu Dhakumin (Fort) | بني ذكومين | Banī Dhakūmīn |
| Banu Ghasan (Fort) | بني غسان | Banī Ghasān |
| Banu Hasan (Fort) | بني حسن | Banī Ḥasan |

| Our Translation | Arabic Name | IJMES Transliteration |
|---|---|---|
| Banu Khatab (Fort) | بني خطاب | Banī Khatāb |
| Banu Måmun (Fort) | بني مأمون | Banī Måmūn |
| Banu Wajas (Port) | بني وجاص | Banī Wajās |
| Banu Walul (Fort) | بني ولول | Banī Walūl |
| Bar | أنتبرية/أنتابروا | Antabariyya/Antābrwā |
| Barashk (City) | برشك | Barashk |
| Barcelona | برشلونة | Barshilūna |
| Bari | باري | Bārī |
| Barletta | برلت | Barlit |
| Basento (River) | برنتال | Burintāl |
| Bashnu (Port) | قرطيل باشنو | Qarṭīl Bāshnū |
| Bashu (Island) | باشو | Bāshū |
| Basil (Fort) | الريحانة | al-Rayḥāna |
| Baška (City) | قسطيلصقة | Qasṭīlṣiqa |
| Baswārīyya (Port) | بسوارية | Baswārīyya |
| Baṭlān | بطلان | Baṭlān |
| Battalari (Castle) | بطالري | Baṭālarī |
| Battifarano (City) | بتبران | Batbirān |
| Béja (City) | باجة | Bāja |
| Béjaïa (City) | بجاية الناصرية | Bijaya al-Nāṣiriya |
| Belcastro (Town) | جنقو قسطروا | Jinqū Qasṭrūwā |
| Benevento (City) | بنبنت | Binbint |
| Bilqan (City) | بيلقان | Bīlqān |
| Biograd (City) | دغواطة | Daghwāṭa |
| Bisceglie | بشتالية / بشالية | Bishtāliyya |
| Bisignano | بسنيان | Bisinyān |
| Biskra (City) | بسكرة | Biskra |
| Bitola | بترلة | Betrila |
| Bitonto (City) | بطنت | Baṭnat |
| Bivona | بيبوني | Bībwanī |
| Bizerte (City) | بنزرت | Binzert |
| Blacks (Lands) | السودان | al-Sūdān |

| Our Translation | Arabic Name | IJMES Transliteration |
|---|---|---|
| Book (Fort) | الكتاب | al-Kitāb |
| Bradano (River) | ابرادنو / ابراطنو | Abrādnū / Abrāṭnū |
| Briga (Port) | حجر أبي خليفة | Hajr Abi Khalīfa |
| Brindisi (City) | ابرندس | Abrindis |
| Brolo (Port) | مرسى دالية | Marsa Dālya |
| Brucato (Castle) | بورقاد | Būrqād |
| Brucoli (River) | وادي زيدون | Wādī Zīdūn |
| Bruzzano (Estate) | برصانة | Barṣāna |
| Buccheri (Estate) | بكير | Bukayr |
| Bucchianico | بكلان | Baklān |
| Buje (City) | بوبلة | Būbla |
| Buna Dabūnsī (Castle) | بنة دبونسي | Buna Dabūnsī |
| Buried (Bay) | المدفون | al-Madfūn |
| Buscemi (Fortress/Stronghold) | أبي شامة | Abī Shāma |
| Bussento (River) | بلقسطروا | Balqasṭrwā |
| Butera (Castle) | بثيرة | Buthīra |
| Butrint (City) | بذرنت | Budhrint |
| Caccamo | ققبش | Qaqbash |
| Cagnano Varano | قينان | Qīnān |
| Calabria (Region) | قلورية | Qalurīyya |
| Calatafimi (Castle) | فيمي | Fīmī |
| Calatamauro | مورو | Mūrū |
| Calatrasi (Castle) | طرزي | Ṭarzī |
| Calatubo | اوبي | Awbī |
| Calatubo River | وادي قلعة اوبي | Wādī Qalʿat Awbī |
| Calcasacco (Castle) | قلقاس / فلفاسق | Qilqās/Falfāsiq |
| Calore (River) | قلور | Qalūr |
| Caltabellotta (Castle) | البلوط | al-Balūṭ |
| Caltagirone (Fortress/ Stronghold) | الخنزارية | al-Khinzāriyya |
| Caltanissetta (Fortress/ Stronghold) | ألنساء | al-Nisāʿ |

| Our Translation | Arabic Name | IJMES Transliteration |
|---|---|---|
| Calvera (City) | قلبرية | Qalbriyya |
| Camel Drivers (Well) | الجمالين | al-Jamālīn |
| Camerota | قمراطة | Qamrāṭa |
| Cammarata (Estate) | قمراطة | Qamrāṭa |
| Campania (City) | قنبانية | Qanbāniyya |
| Campo di Giove | منت دجون | Munt Dajūn |
| Campofelice di Roccella (Estate and Port) | صخرة الحديد | Ṣakhrat al-Ḥadīd |
| Campomarino | قنب مارين | Qanb Mārīn |
| Cancello | جنقالة | Jinqāla |
| Candelaro (River) | قنالار | Qanālār |
| Cannae (City) | قاني | Qānī |
| Canosa | قنوصة | Qanūṣa |
| Canosa Sannita | فونصة | Fawniṣa |
| Capaccio | قبواح | Qabwāḥ |
| Capizzi (Castle) | قيسي | Qīsī |
| Capo Calavà (Port) | رأس خلي | Rās Khalī |
| Capo di Gallo (Port) | غالة | Ghāla |
| Capo di Mascali (Port) | قرطيل مسقلة | Qarṭīl Masqala |
| Capo di Milazzo (Port) | الرأس | al-Rās |
| Capo di Orlando (Port) | جفلودي الصغرى | Jiflūdi al-Ṣughayra |
| Capo di Rama (Port) | برطنيق | Barṭanīq |
| Capo di Tindari (Port) | رأس دنداري | Rās Dindārī |
| Capo Feto (Port) | الرأس | al-Rās |
| Capo Granitola (Port) | رأس البلاط | Rās al-Balāṭ |
| Capo Molini (Port) | وادي لياج | Wādī Liāj |
| Capo Murro di Porco (Port) | أنف الخنزير | Anf al-Khanzīr |
| Capo Passero | مرسى البوالص | Marsa al-Bawāliṣ |
| Capo Pellaro | رأس فلامة | Rās Filāma |
| Capo Peloro (Port) | الفارو | al-Fārū |
| Capo Plaia (Port) | أنف الآخر | Anf al-Ākhar |
| Capo Raisgerbi (Port) | أنف الكلب | Anf al-Kalb |

| Our Translation | Arabic Name | IJMES Transliteration |
|---|---|---|
| Capo San Marco (Port) | أنف النسر | Anf al-Nasr |
| Capo Santa Panagia (Port) | خندق الغريق | Khandaq al-Gharīq |
| Capo Schisò (Port) | القصوص | al-Qaṣūṣ |
| Capua (City) | قبوة | Qabwa |
| Capraia (Island) | قبريرة | Qabrīra |
| Capri (Island) | قبرة | Qabra |
| Carabo (River) | وادي القارب | Wādī al-Qārib |
| Carbone | قربون | Qarbūn |
| Carcassonne | قرقسونة | Qarqasūna |
| Carceri | قرجرة | Qarjara |
| Carini (Town) | قرنيش | Qarnīsh |
| Caronia (Castle) | القارونية | al-Qārūniyya |
| Carthage (City) | قرطاجنة | Qarṭājina |
| Carunchio (Castle) | قارنقل | Qārnaqal |
| Caselle (Town and River) | قزلو / قزله | Qazlū / Qazluh |
| Cassibile (Estate) | قسباري | Qasbāri |
| Castel del Giudice | قشطة جرزد | Qishṭa Jarzid |
| Castel di Iudica (Estate) | جوذقة | Jūdhiqa |
| Castel di Jonato | مكشتر يناطة | Makshatar Yināṭa |
| Castel di Sangro | صنغرة | Ṣanghara |
| Castel Missanello (Castle) | قشتال / قشتال مشتال | Qishtāl / Qishtāl Mishtāl |
| Castellamare del Golfo (Castle and Port) | المدارج | al-Madārij |
| Castellamare di Velia | قشتال دمار | Qashtal Damār |
| Castello Mano | قشتال مان | Qishtāl Mān |
| Castelnuovo | قشتال نوبة / قشتال نوب | Qishtāl Nūba / Qishtāl Nūb |
| Castiglione (Castle) | قسطلون | Qasṭalūn |
| Castiglione Messer Marino | قستلنس/قشطلون | Qastilnas / Qashṭilūn |
| Castro (Castle) | قاسطروا | Qāsṭirwā |
| Castro (City) | قشطرة | Qashṭara |
| Castrocucco | قسطر كلي/قشطرقفلو | Qasṭr Kalī/Qashṭr Qaflū |
| Castronovo | قصر نوبو | Qaṣr Nūbū |

| Our Translation | Arabic Name | IJMES Transliteration |
|---|---|---|
| Castronuovo di Sant'Andrea (Region) | قسطر نوب | Qasṭar Nūb |
| Castrovillari | قطروبلي | Qaṭrūblī |
| Catania (City) | قطانية | Qaṭāniyya |
| Catanzaro (Castle) | قطنسان / قطنصار | Qaṭansān / Qaṭanṣār |
| Cefalù (City) | جفلودي | Jiflūdī |
| Celenza Valfortore (City) | جرسنة | Jarsana |
| Cephalonia (Island) | جفلونية | Jiflūniyya |
| Ceppaloni (Castle) | جبرون | Jabrūn |
| Cerami (Estate) | جرامي | Jirāmī |
| Cerenzia | جرنتية / جرنسية | Jarnitiyya / Jarinsiyya |
| Cerignola | غرنيلية | Gharnīliyya |
| Cerro (Castle) | جروا | Jarwā |
| Cervaro (River) | كاطة | Kāṭa |
| Ceuta (City) | سبتة | Sebta |
| Cherchell | شرشال | Shershāl |
| Chiaramonte | كلرمنت / أكلرمنت | Kalrimint / Aklirimint |
| Chieti | تات | Tāt |
| Church (Mountain Edge) | الكنيسة | al-Kanīsa |
| Cimitile | جيبطيرة | Jībṭīra |
| Cimitile | جبيطيرة | Jibīṭīra |
| Cinisi (Estate) | شنس | Shins |
| Cintursi | منت طري | Munt Ṭarī |
| Circeo | جرنجوا / جرجير / قيطنة العرب | Jarnjwā / Jarjīr / Qayṭanat al-ʿarab |
| Cirella (Cape and River) | رأس جرلة | Rās Jarla |
| Cirò | إبسخرو | Ibsakhrū |
| Civitaluparella (Castle) | جبيطة لبرال | Jabīṭa Librāl |
| Civitate | جبطاط / جبطات | Jibṭāṭ / Jibṭāt |
| Child (Port) | جرف الطفل | Jarf al-Ṭifl |
| Cliff (Fort) | الجرف | al-Jurf |
| Colle Archiano | أرقلان | Arqilān |
| Collesano (Stronghold/Fortress) | الصراط | al-Ṣirāṭ |

| Our Translation | Arabic Name | IJMES Transliteration |
|---|---|---|
| Colimena (Port) | قيطولقة | Qīṭūlqa |
| Colobraro (Castle) | قلبراط | Qalbrāṭ |
| Column of the Shipmasts | افلومية السواري | Iflūmiyya al-Sawārī |
| Comino (Island) | كمونة | Kamūna |
| Constantine | قسنطينة | Qasnaṭīna |
| Conversano (City) | قنبرصان | Qanbirṣān |
| Conza della Campania | قنس | Qans |
| Corato (City) | قورات | Qūrāt |
| Corbella | قربلة | Qarbala |
| Corfu (Island) | قرفس | Qarfas |
| Corleone (Castle) | قرليون | Qurlīyyūn |
| Corsica (Island) | قرشقة | Qurshiqa |
| Cres (Island) | جرسة | Jarsa |
| Croatia (Region) | جرواسية | Jurwāsiyya |
| Crotone (City) | قطرونة / قطروني | Qaṭrūna / Qaṭrūnī |
| Cuma (City) | كومة | Kūma |
| Cyclopean Isles (Port) | جزائر لياج | Jazāʿir Liāj |
| Dahmani (City) | ابة | Āba |
| Dakama (Village) | دكمة | Dakama |
| Date Palm (Fort) | النخلة | al-Nakhla |
| Date Palm (Fort) | النخيل | al-Nakhīl |
| Degache (City) | تقيوس | Taqiyus |
| Demmer (Mountain) | دمر | Damar |
| Devia (Region) | دابية | Dābiyya |
| Diano | أديانة | Adyāna |
| Djerba (Island) | جربة | Jerba |
| Dogs (Swamp) | سباخ | al-Kilāb |
| Donkey's Backbone (Bay) | صلب الحمار | Ṣulb al-Ḥimār |
| Dragonara | أذرغونالة | Adhraghūnāla |
| Drinpoli | أذرنوبلي | Adhranūblī |
| Dubrovnik (City) | رغوص / رغوصة | Raghūṣ / Raghūṣa |
| Dungeon | الديماس | al-Dīmās |

| Our Translation | Arabic Name | IJMES Transliteration |
|---|---|---|
| Đurinići (City) | دلجينة | Duljīna |
| Durrës (City) | دراست/ادراست | Durāsit/ Adrāst |
| Earring of Maria (Port) | قرط مارية | Qurt Māriyya |
| Eboli | ابلة | Abla |
| Edough (Mountain) | يدوغ | Īdūgh |
| Elba (Island) | البة | Alba |
| El Hamma du Jérid (City) | الحمّة | al-Ḥamma |
| El Hamma du Jérid (City) | الحمة | al-Ḥamma |
| El Hammadia (Village) | اوسحنت | Awsaḥanat |
| El Hammamet | الحمامات | al-Ḥamāmāt |
| El Kala (City) | مرسى الخرز | Marsa al-Kharaz |
| Elevation (Fort) | العالية | al-ʿĀliya |
| Enna (City) | قصرياني | Qaṣryānī |
| Entrance of the Bowls (Port) | دخلة القصاع | Dakhlat al-Qaṣāʿ |
| Erice (Mountain) | جبل حامد | Jabal Ḥāmid |
| Falconara (Port) | مرسى الشلوك | Marsa al-Shalūk |
| Fallo (City) | الفلفال | al-Filfāl |
| Fāmyū | فاميو | Fāmyū |
| Fara di Pescara | فارة بشكارة | Fāra Bishkāra |
| Fara San Martino (Castle) | فارة | Fāra |
| Favale | فياض | Fayāḍ |
| Favignana (Island) | الراهب | al-Rāhib |
| Filicudi (Island) | فيكوذة | Fīkūdha |
| Fiume della Verdura (River) | وادي البو | Wādī Albū |
| Fiume di Alcantara (River) | الوادي البارد | al-Wādī al-Bārid |
| Fiume di Cassibile (River) | وادي قسباري | Wādī Qasbārī |
| Fiume di Castellamare (River) | وادي المدارج | Wādī al-Madāraj |
| Fiume di Imera (River) | وادي السواري | Wādī al-Sawārī |
| Fiume di Misilmeri (River) | وادي الامير | Wādī al-Amīr |
| Fiume di Ragusa (River) | وادي رغوص | Wādī Raghūṣ |
| Fiume di Salso (River) | الوادي الملح | al-Wādī al-Milḥ |
| Fiume di Saponara (River) | وادي عبود | Wādī ʿAbūd |

| Our Translation | Arabic Name | IJMES Transliteration |
|---|---|---|
| Fiume di Torto (River) | وادي ابي رقاد | Wādī Abi Raqqād |
| Fiume Ippari (River) | كرني | Karnī |
| Foggia | فوج | Fawj |
| Forca (Castle) | فركة | Furka |
| Foro (River) | لوقدو | Lūqdū |
| Fortore (River) | فرطول | Furṭūl |
| Francavilla Angitola (Castle) | أنجيطلوا | Anjīṭalwā |
| Frigento | فراجنطو | Frājinṭū |
| Gabès (City) | قابس | Qābis |
| Gaeta (City) | غيطة | Ghīṭa |
| Gafsa (City) | قفصة | Qafṣa |
| Gagliano | غليانية | Ghalyāniyya |
| Gagliano Aterno (Castle) | بليان | Balyān |
| Galati Mamertino (Stronghold/Fortress) | غالط | Ghalāṭ |
| Gallicchio | قابلي | Qāblī |
| Gallipoli (City) | قليبلي | Qalīblī |
| Ganano | غنانو | Ghanānū |
| Garigliano (River) | غرليان | Gharlyān |
| Gascony (Region) | غشكونية | Ghashkūniyya |
| Gerace (City and River) | جراجي | Jerājī |
| Geraci Siculo | جاراس | Jārās |
| Geyser (Stronghold/Fortress) | الحمة | al-Ḥamma |
| Ghana (Lands) | غانة | Ghāna |
| Gharghara (Fort) | غرغرة | Gharghara |
| Giglio (Island) | اليالية | Al-Yālya |
| Giovinazzo | جبناس | Jibnās |
| Girone | جرندة | Jiranda |
| Gissi (Castle) | جنس | Juns |
| Giuliano (Rock) | رقة فنجلان | Ruqqa Fanjalān |
| Godrano | الغدران | al-Ghudrān |
| Gozo (Island) | غودش | Ghawdash |

| Our Translation | Arabic Name | IJMES Transliteration |
|---|---|---|
| Granary (Fort) | الهرى | al-Hara |
| Gratteri (Stronghold/Fortress) | قرطيرش | Qarṭīrsh |
| Gravina (City) | غرابينة / اغربينة | Ghrābīna / Ighribīna |
| Greek (River and Port) | وادي إغريقو | Wādī Ighrīqū |
| Grottole (City) | أغرطلو / اكرت | Aghraṭlū / Akrit |
| Grumento Nova | صبونارة | Ṣabūnāra |
| Guaceto / Torre Guaceto (Islands) | غوشيت | Ghūshīt |
| Guardiagrele | غردية | Ghardiyya |
| Guardialfiera | غاردية ارط | Ghārdiyya Arṭ |
| Gulf of the Two Rivers | جون الواديين | Jūn al-Wādyīn |
| Habala (Fort) | حبلة | Ḥabala |
| Hashim (Fort) | هاشم | Hāshim |
| Hawwara (Village) | هوارة | Hawwara |
| Hergla (Castle) | هرقلية | Heraqliya |
| Hergla (Castle) | اهرقلية | Aharaqliya |
| Himarë (City) | جمارة | Jamāra |
| Himarë (City) | جمارة / برنتو | Jamāra / Barnatū |
| Hunter (Fort) | صياد | Sayad |
| Ianina | يانئة | Yānana |
| Ibn al-Ja'd (Fort) | ابن الجعد | Ibn al-Ja'd |
| Ibn Mathkud (Little Market) | ابن مثكود | Ibn Mathkūd |
| Ibn 'Ayshun (Fort) | ابن عيشون | Ibn 'Ayshūn |
| Ifran | افران | Ifrān |
| Inland Melfi | ملف البرية | Milf Bāriyya |
| Irsina (City) | منت فنجوس | Munt Fanjūs |
| Ischia (Island) | شكلة | Shikla |
| Island of the Dove (Port) | جزيرة الحمام | Jazīrat al-Ḥammām |
| Isma'il (House) | إسماعيل | Ismā'īl |
| Isnello | الحمار | al-Ḥimār |
| Isola dei Porri (Port) | جزيرة الكراث | Jazīrat al-Karāth |
| Isola delle Femmine (Port) | الجزيرة | al-Jazīra |
| Isola di Capo Rizzuto (Island) | اوسلة | Awsala |

| Our Translation | Arabic Name | IJMES Transliteration |
|---|---|---|
| Istria (Region) | استرية | Astriyya |
| Izola | انملة / انجلة | Inmala / Injala |
| Jaham (Fort) | جهم | Jaham |
| Jamunas (City) | جمونس | Jamūnas |
| Jardan (Fort) | جردان | Jardān |
| Jarjis (City) | جرجيس | Jarjīs |
| Jarjis (Fort) | جرجيس | Jarjīs |
| Jato (Castle) | جاطوا | Jāṭwā |
| Jifla (Town) | جفلة | Jifla |
| Jijel (City) | الشامي | al-Shāmī |
| Jilla (Fort) | جلة | Jila |
| Kaboudia (Fort) | قبوذية | Qabūdhiya |
| Kairouan (City) | القيروان | al-Qayrawan |
| Karin (Cape) | كرين | Karīn |
| Kelibia (Castle) | اقليبية | Iqlībiya |
| Khabat (Wells) | خبت | Khabat |
| Kīra | كيرة | Kīra |
| Koper (Town) | دستريس | Dastarīs |
| Korba (Fort) | قربة | Qorba |
| Korbous (Fort) | قربص | Qurbus |
| Kotor (City) | قاترة/قاطروا / قاذروا | Qātra/Qāṭruwā / Qādhruwā |
| Kugha | كوغة | Kūgha |
| Kuriat (Island) | قورية | Qūriya |
| Kutin (Fort) | كوطين | Kūṭīn |
| Labin (City) | درونة/البونة | Darūna/Albūna |
| Ladis (River) | لادس | Lādis |
| Lagopesole (Town) | لقبارة | Liqbāra |
| Lama dei Peligni | لامة | Lāma |
| Lampedusa (Island) | لنبذوشة | Lanbadhūsha |
| Lampione (Island) | جزيرة الكتاب | Jazīrat al-Kitāb |
| Lamta (Fort and Island) | لمطة | Lamṭa |
| Lanciano | أنكزمة | Ankazma |

| Our Translation | Arabic Name | IJMES Transliteration |
|---|---|---|
| Lantishak (Castle) | لنتشك | Lantishak |
| Lao (River) | لاينة | Lāyna |
| Le Castella | قشتال | Qashtāl |
| Lebda (City) | لبدة | Lebda |
| Lebna (Fort) | لبنة | Lebna |
| Lecce (City) | لج | Lij |
| Lefkada (Island) | لقاطة | Liqāṭa |
| Lentini (Stronghold/Fortress) | لنتيني | Lintīnī |
| Lesina (City) | لاشنة / لازنة / لشنة | Lāshna / Lāzna / Lashna |
| Leuca (Cape) | ليغة | Līgha |
| Levanzo (Island) | اليابسة | al-Yābisa |
| Licata (Castle) | لنبياذة | Linbiyādha |
| Licosa (Island) | بغوضة | Baghūḍa |
| Linosa (Island) | نموشة | Namūsha |
| Lipari (Island) | ليبر | Lībar |
| Liscia (Town) | لشيوش | Lishyūsh |
| Livestock (House) | الدواب | al-Dawab |
| Locadi (Stronghold/Fortress) | لوغاري | Lūghārī |
| Lopar (City) | لوبارة | Lūbāra |
| Lorbeus (City) | الاربس | al-Arbus |
| Losinj (Island) | أوسر | Awser |
| Lovran (City) | الاورنة | Alawrna |
| Luhaqa (City) | لوحقة | Lūhaqa |
| Lūrant | لورنت | Lūrant |
| M'Sila (City) | المسيلة | al-Masīla |
| Macchia | ماقلة | Māqila |
| Magi (Fort) | المجوس | al-Majūs |
| Mahdia (City) | المهدية | al-Mahdiya |
| Majana (City) | مجانة | Majāna |
| Majdunis (Fort) | مجدونس | Majdūnis |
| Malta (Island) | مالطة | Mālṭa |
| Manfria (Port) | مرسى بثيرة | Marsa Bithīra |

| Our Translation | Arabic Name | IJMES Transliteration |
|---|---|---|
| Maniace (Village) | منياج/ غيران الدقيق | Minyāj / Ghīrān al-Daqīq |
| Manorca (Island) | منورقة | Manūrqa |
| Mansouria (City) | صبرة | Ṣabra |
| Maqada (Mountain) | مقدة | Maqada |
| Maratea | مراثيا | Mirāthyā |
| Marettimo (Island) | مليطمة | Milīṭma |
| Màrgana (Estate) | مرغنة | Marghana |
| Marikya (Port and Fort) | مركيا | Markiya |
| Marina di Fiumdinisi (Port) | الأجاصة | al-Ajāṣa |
| Marina di Ragusa (Port) | غدير الشرشور | Ghadīr al-Sharshūr |
| Marineo | مرناو | Marnāw |
| Marmajana (City) | مرماجنة | Marmājana |
| Marsala (City) | مرسى علي | Marsa ʿAlī |
| Marsico | مرسقو | Marsuqū |
| Marsicovetere | مرسقة القديمة | Marsiqa al-Qadīma |
| Martirano | مرطران | Marṭirān |
| Marzamemi (Port) | مرسى الحمام | Marsa al-Ḥammām |
| Mascali (Village) | مصقلة | Maṣqala |
| Matera (City) | متيرة | Matīra |
| Materada (City) | طامطرس | Ṭāmṭaras |
| Mattinata | ماتناطة | Mātnāṭa |
| Mazara (City) | مازر | Māzar |
| Mediterranean (Sea) | الجزء الثاني | al-Baḥr al-Shāmī |
| Medjerda | بجردة | Bejerda |
| Medulin (City) | موذولينة | Mūdhūlīna |
| Melito (River) | وادي العسل | Wādī al-ʿAsal |
| Melloulèche (Fort) | مليان | Maliyan |
| Mercuri Tedesco | مركوري | Markūrī |
| Meskiana (Village) | مسكيانة | Miskiyana |
| Messina (City) | مسينى | Misīnā |
| Mezzojuso | منزل يوسف | Manzil Yūsuf |
| Middle Step (Port) | الدرجة الوسطى | al-Daraja al-Wasṭā |

| Our Translation | Arabic Name | IJMES Transliteration |
|---|---|---|
| Miglionico (Town) | مليون | Milyūn |
| Milazzo (Castle) | ميلاص | Mīlāṣ |
| Mills of Taranto (River) | مطاحن طارنت | Maṭāḥin Ṭārant |
| Mineo (Stronghold/Fortress) |ميناو | Mīnaw |
| Mingardo (River) | شنت سيمري | Shant Sīmrī |
| Mingardo (River) | مولبة | Mawlaba |
| Minzit (Fort) | مينزت | Mīnzit |
| Mirabella Eclano (Castle) | اقترندة | Iqtaranda |
| Miseno (Port) | مسينة | Masīna |
| Misilindino (Estate) | منزل سندي | Manzil Sindī |
| Misilmeri (Castle) | الأمير | al-Amīr |
| Mistretta (Castle) | قلعة القوارب | Qalʿat al-Qawārib |
| Mīyūr | ميور | Mīyūr |
| Modica | موذقة | Mūdhiqa |
| Molfetta | ملبنت / ملفنت | Malbant / Malfant |
| Molva | مولبة | Mūlba |
| Mondello (Port) | مرسى الطين | Marsa al-Ṭīn |
| Monopoli (City) | منوبلي | Manūblī |
| Montalbano Elicona (Stronghold/Fortress) | منت البان | Munt Albān |
| Monte Barello (Mountain) | أولاد برال | Awlād Barāl |
| Monte Maiella (Mountain) | مايال | Māyāl |
| Monte Melo | منت مللو | Munt Millū |
| Monte Rossello (Port) | الأختين | al-Ukhtīn |
| Monte Sant'Angelo | شنت أنجلي/ شنت أنجلو | Shant Anjlī / Shant Anjlū |
| Monte Terminio (Mountain) | منتال | Mantāl |
| Montebello sul Sangro (Castle) | منت بال | Munt Bāl |
| Montecalvo Irpino | منت قلوي | Munt Qalwī |
| Monteferrante (Castle) | منت افرند / منت فرن | Munt Ifrand / Munt Frand |
| Monteforte Cilento | منت فرت | Munt Furt |
| Monteforte San Giorgio | منت دفرت | Munt Dafart |
| Montefosco (Castle) | منت فسق | Munt Fasq |

| Our Translation | Arabic Name | IJMES Transliteration |
|---|---|---|
| Montefredano (Mountain) | ابراطنة | Ibrāṭna |
| Montemitro (Castle) | منت دمروا | Munt Damrwā |
| Montemurro | منت مور / منت ميور | Munt Mūr / Munt Miyūr |
| Montesano | منت صان | Munt Ṣān |
| Montesarchio (City) | منت سرح | Munt Sarḥ |
| Montoro | منت ابوو | Munt Abwū |
| Motta Sant'Anastasia | نسطاسية | Nasṭāsiyya |
| Mottola | موطلى / ماتلى / ماطلى / ماتي | Mūṭlā/ Mātlā / Māṭlā / Mātī |
| Mucata (River) | مخاطة | Makhāṭa |
| Nabeul (Fort) | نابل | Nābil |
| Nafusa (Mountain) | نفوسة | Nafusa |
| Naples (City) | نابل | Nābil |
| Naqawas (City) | نقاوس | Naqawas |
| Narbonne | اربونة | Arbūna |
| Nardo (Port) | نارطس | Nārṭas |
| Naro (Estate) | ناروا | Nārwā |
| Naso (Castle) | ناصو | Nāṣū |
| Nefta (City) | نفطة | Nafṭa |
| Nefzaoua (City) | نفزاوة | Nefzawa |
| Neto (River) | نيطو | Nīṭū |
| Nicosia (Castle and River) | النيقشين | al-Nīqishīn |
| Nicotera | نقوطرة | Naqūṭra |
| Nin (City) | نونة / نينص | Nūna / Nīnṣ |
| Nona (City) | نولنص | Nūlinṣ |
| Noto (Stronghold/Fortress) | نوطس | Nūṭas |
| Novrigrad (City) | جبطنوبة | Jabṭanūba |
| Nuba (Fort) | نوبة | Nūba |
| Nūblū (City) | نوبلو | Nūblū |
| Numana | اقاما/قامة | Iqāma/Qāma |
| Numana (City) | موقة | Mūqa |
| Ofanto (River) | لوذرة | Lūdhra |

| Our Translation | Arabic Name | IJMES Transliteration |
|---|---|---|
| Ognina (Port) | الانكنة | al-Inkana |
| Oliva (River) | أولبة | Awbla |
| Oliveri (Estate and Port) | لبيري | Lībīrī |
| Omišalj (City) | مصقلة/مصقلة | Miṣqila/Maṣqala |
| Ongnina (Port) | الأنكنة | al-Ankana |
| Oppido Lucano (City) | لبوتراي | Libūtrāy |
| Ortona (City) | أطرونة /أرطونة | Aṭrūna / Arṭūna |
| Ostuni | اسلمونة | Islamūna |
| Othonoi (Island) | ثنو | Thanū |
| Otranto (City) | أذرنت | Adhrant |
| Ouargla (City) | وارقلان | Wārqalān |
| Ouasselat (Mountain) | واسلات | Wasilāt |
| Pacentro (Castle) | باجنرة / بجنبرو | Bājnira/Bajinbrū |
| Pacile (Castle) | باشلة | Bāshila |
| Padula | بذولة | Badhūla |
| Pag (Island) | باغا | Bāghā |
| Palazzolo Acrèide | بلنسول | Balansūl |
| Palena (Castle) | بلانة / بالانة | Bilāna/Bālāna |
| Palermo (City) | بلرم | Balarm |
| Palma Campania | بالمة | Bālma |
| Palmarola (Island) | مونسة | Mūnsa |
| Palmoli | بالملة | Bālmala |
| Pantalica | بنتارغة | Bantārgha |
| Pantelleria (Island) | قوصرة | Qūṣra |
| Parga (City) | فاسكيو | Fāskiyyū |
| Partinico (Region) | برطنيق | Barṭanīq |
| Paternò | باترنو | Bātirnū |
| Patria (Village and Port) | بطرية | Baṭriyya |
| Pattarriti | بثرة | Bithra |
| Patti (Castle and Port) | بقطش | Baqṭash |
| Penisola Magnisi (Port) | جزيرة مسمار | Jazīrat Masmār |
| Pennadomo | بنة تاتلي / توتلي | Buna Tātlī / Tūtlī |

| Our Translation | Arabic Name | IJMES Transliteration |
|---|---|---|
| Pennapiedimonte (Castle) | بنة انكي | Bina Ankī |
| Pentedattilo | بنتدقطلة | Bantadaqṭila |
| Pescara (City) | بشكار | Bishkār |
| Pescasseroli | بشكة سارلة | Bishka Sārla |
| Peschici | بسكيش | Biskīsh |
| Petracucca (River) | بطرقوقة | Baṭraqūqa |
| Petralia Sottana (Castle) | بطرلية | Baṭraliyya |
| Petrosa | اطربس | Aṭrabas |
| Petterana (Castle) | بثرانة | Bithrāna |
| Piazza Armerina | ابالطسة | Iblāṭsa |
| Pianosa (Island) | بانوسة | Bānūsa |
| Pietra Padella | حجر ابن الفتى | Hajr Ibn al-Fatā |
| Pietrabbondante | باترة أندنت | Bātra Andant |
| Pietrapaola | باتربول | Bātrubūl |
| Pietraperzia (Stronghold/Fortress) | الحجر مثقوب | al-Ḥajjar al-Mathqūb |
| Platani (Castle and River) | ابلاطنو | Iblāṭnū |
| Plomin | فلامونة | Filāmūna |
| Policastro (Cape and River) | بلقسطروا | Balqasṭrūa |
| Policastro (Castle) | بلي قشطرو | Balī Qashṭrū |
| Policoro (Castle) | بلقوري | Balqūrī |
| Polignano (Castle) | بلنيان | Balinyān |
| Polizzi (Castle) | بولس | Būlis |
| Polla | أبلة | Abla |
| Pomarico Vecchio (Town) | ترتغرير | Tartighrīr |
| Ponza (Island) | بونسة | Būnsa |
| Poreč (City) | برنجوا / برنزوا | Baranjwā / Barnazwā |
| Port of the Dromon (Port) | مرسى الدرامن | Marsa al-Darmān |
| Port of the Expert (Port) | مرسى الحذاق | Marsa al-Ḥadhāq |
| Port of the River (Fort) | مرسى الوادي | Marsa al-Wadī |
| Port of the Tree (Port) | مرسى الشجرة | Marsa al-Shajara |
| Porto di Ulisse/La Marza (Port) | مرسى البولص | Marsa al-Bawāliṣ |

| Our Translation | Arabic Name | IJMES Transliteration |
|---|---|---|
| Porto Palo | ترسة أبي ثور/ كرم الرنبوح | Tarsat Abi Thūr/Karam al-Runbūḥ |
| Porto Xifonio (Port) | أكسيفو | Aksīfū |
| Positano (Port) | بسطانة | Basṭāna |
| Potenza (City) | بتانسة / بتنسية | Bitānsa / Bitinsiyya |
| Pozzuoli (Castle) | قشتلي | Qashtly |
| Prata (Castle) | برات | Barāt |
| Pretoro (Castle) | براتور | Barātūr |
| Prizzi (Castle) | برزوا | Barzwā |
| Pula | بولة | Būla |
| Punta Campanella (Cape) | رأس منتيرة | Rās Mantīra |
| Qalayusha (Cape) | قاليوشا | Qāliyushā |
| Qamuda | قمودة | Qamūda |
| Qanan (Cape) | قانان | Qānān |
| Qanata (Fort) | قناطة | Qanāṭa |
| Qandīla | قنديلة | Qandīla |
| Qarsilūn (Mountain) | قرسلون | Qarsilūn |
| Qasara (City) | قاصرة | Qāṣara |
| Qasas (Fort) | قاساس | Qāsās |
| Qastiliya (City) | قسطيلية | Qasṭīliya |
| Qastilnas (Mountain) | قستلنس | Qastilnas |
| Qatrashāl (Mountain) | قترشال | Qatrashāl |
| Qazil (Fort) | قزل | Qazil |
| Qishtāl Lūrint | قشتال لورنت | Qishtāl Lūrint |
| Rab (Island) | أربا | Arbā |
| Rabah (Spring) | رباح | Rabāḥ |
| Raghugha | رغوغا | Raghūghā |
| Ragusa (Stronghold/Fortress) | رغوص | Raghūṣ |
| Rālbinū | رالبنو | Rālbinū |
| Randazzo (Village) | الرندج | al-Randaj |
| Rapino | رتينة | Ratīna |
| Raqqada (Forts) | رقادة | Raqāda |

| Our Translation | Arabic Name | IJMES Transliteration |
|---|---|---|
| Reggio (City) | ريو | Rīyū |
| Rība | ريبة | Rība |
| Ripalta | ربالد | Ribāld |
| Rivello (River) | الربل | al-Rabl |
| River of the Masts (River) | وادي السواري | Wādī al-Sūwārī |
| Rivers (Cape) | الاودية | al-Awdiya |
| Rivolo (River) | ريغلو / نيقلو | Rīghlū / Nīqlū |
| Roca Vecchia | شوذة | Shūdha |
| Rocca del Abate | بتسي | Bitsī |
| Rocca Giuliopoli | ديبلوا | Dīblwā |
| Rocca Gloriosa | اكلريوزة | Iklaryūza |
| Rocca Nova | رقة فيلب | Ruqqa Fīlib |
| Rocca Qūrālī | رقة قورالي | Rocca Qūrālī |
| Roccabascerana | رقة بلتة | Ruqqa Balta |
| Roccamontepiano | رقة منت بلان | Ruqqa Munt Balān |
| Roccella (Port) | الصخرة | al-Ṣakhra |
| Rodi Garganico | روذانة / روذنة | Rūdhāna / Rūdhna |
| Rofrano | رفران | Rafrān |
| Rogoznica (City) | واوغوري / لوغارو | Wawghūry / Lūghārū |
| Rome (City) | رومة | Rūma |
| Roseto | روسيت | Rūsīt |
| Rossano | رشانة / رسيانو | Rashāna / Rasyānū |
| Rovinj (City) | ريغنو | Rīghnū |
| Rum (Forts) | روم | Rūm |
| Ruqqa Shanj | رقة شنج | Ruqqa Shanj |
| Rūthmart (Settlement) | روثمرت | Rūthmart |
| Ruvo (City) | روبة | Rūba |
| Sabato (River) | شابطو | Shābṭū |
| Sabratha (Fort) | صبرة | Ṣabra |
| Saint Nicholas (Port) | شنت نقولة | Shant Niqūla |
| Saja (Fort) | سجة | Saja |
| Sala Consilina | الصالة | al-Ṣāla |

| Our Translation | Arabic Name | IJMES Transliteration |
|---|---|---|
| Salakta (Fort) | سلقطة | Salaqṭa |
| Salapia (City) | صلبي | Ṣalbī |
| Salerno (City) | سلرنو | Salarnū |
| Salih (Fort) | صالح | Sāliḥ |
| Salina (Island) | دندمة | Dandama |
| Saltworks (Port) | الملاحة | al-Milāḥa |
| Sambiase | شنت بجوش | Shant Bijūsh |
| Samiya (Fort) | سامية | Sāmiya |
| San Barbato | شنت برباط | Shant Barbāṭ |
| San Brancato | بنس ادرات | Bins Adrāt |
| San Chirico Raparo | شنت اكلركو | Shant Iklarkū |
| San Donato Val di Comino | شنت دناط | Shant Dināṭ |
| San Felice (Port) | شنت فليجي | Shant Filījī |
| San Fratello | فيلادنت | Fīlādint |
| San Leonardo (River) | وادي لنتيني | Wādī Lentīnī |
| San Lorenzo | شنت لورين / شنت لورنس | Shant Lūrīn / Shant Lūrins |
| San Marco d'Alunzio (Stronghold/Fortress) | شنت مركو | Shant Markū |
| San Martino di Agri | شنت مرتين | Shant Martīn |
| San Martino di Agri (City) | أكري | Akrī |
| San Mauro | شنت موروا | Shant Mūrwā |
| San Severo | شنت صبير | Shant Ṣabīr |
| San Teodoro Vecchio (Church) | شنت توذر | Shant Tūdhar |
| San Vitale | شنت فيلس | Shant Fīls |
| San Vito (Port) | شنت بيطو | Shant Bīṭū |
| Ṣanāla | صنالة | Ṣanāla |
| Sand (Cape) | الرملة | al-Ramla |
| Sant' Eufamia | شنت فيمي | Shant Fīmī |
| Sant'Agata de'Goti | شنت غاثي | Shant Ghāthī |
| Sant'Agata di Puglia | رقة شنت غاثي | Ruqqa Shant Ghāthī |
| Sant'Andrea | قسطرنوب | Qasṭarnūb |
| Sant'Anna (Estate) | رحل قنطانة | Raḥl Qanṭāna |

| Our Translation | Arabic Name | IJMES Transliteration |
|---|---|---|
| Sant'Arcangelo (City) | شنت أركنجل | Shant Arkanjal |
| Sant'Eufamia | شنت فيمة | Shant Fīma |
| Santa Croce (Port) | رأس الصليبة | Ras al-Ṣalība |
| Santa Maria | شنت مارية | Shant Māriyya |
| Santa Maria al Bagno (Port) | طراجة | Ṭarāja |
| Santa Severina | شنت سميرى | Shant Samīra |
| Santa Tecla (Port) | شنت شقلى | Shant Shaqla |
| Santo Stefano di Briga (Port) | شنت اصطيفن | Shant Iṣṭīfin |
| Santo Vito (Port) | جبل شنت بيطو | Jabal Shant Bīṭū |
| Sarba (Fort) | سربة | Sarba |
| Sarconi | سرقونة / سرقون | Sarqūna / Sarqūn |
| Sardinia (Island) | سردانية | Sardāniyya |
| Sarno | سرنة | Sarna |
| Satfura (City) | سطفورة | Saṭfūra |
| Savone (River) | جلاج | Jalāj |
| Saw (Beach) | المنشار | al-Minshār |
| Saʿd (Fort) | سعد | Saʿd |
| Sbeitla (City) | سبيطلة | Sabīṭala |
| Sbiba (City) | سبيبة | Sabība |
| Scalea | دسقالية | Dasqāliyya |
| Scalea (Castle) | دسقالية | Disqāliyya |
| Scaletta Zanklia (Port) | الدرجة الصغيرة | al-Darajat al-Ṣughayra |
| Scanno | أسقنو | Isqunū |
| Scanza | شنت صيتري | Shant Ṣītrī |
| Sciacca (Town) | الشاقة | al-Shāqa |
| Scicli (Stronghold/Fortress and Port) | شكلة | Shakla |
| Sclafani Bagni | سقلفية | Saqlāfiyya |
| Sele (River) | اسيلو | Asīlū |
| Sele (River) | السيلو | Sīlū |
| Senise (Town and River) | سينس / سنيس | Sīns / Sinīs |
| Senj | آسيا/سينة | Āsyā/Sīna |

| Our Translation | Arabic Name | IJMES Transliteration |
|---|---|---|
| Sentry (Fort) | المرصد | al-Marṣad |
| Serino | سرين | Sirīn |
| Sessa Aurunca (River) | ساسة | Sāsa |
| Sfax (City) | سفاقس | Sfāqus |
| Shamakh (Fort) | شماخ | Shamākh |
| Shant Bālmī (Port) | شنت بالمي | Shant Bālmī |
| Shant Bārid Kamīra | شنت بارد كميرة | Shant Bārid Kamīra |
| Shant Juwān Martūblī (Village) | شنت جوان مرتوبلى | Shant Juwān Martūblī |
| Shant Jūwān Mayūr | شنت جوان ميور | Shant Jūwān Mayūr |
| Sharikas (Fort) | شريكس | Sharīkas |
| Sharus (City) | شروس | Sharūs |
| Shelter/Refuge of Khalil | ملجأ خليل | Maljā' Khalīl |
| Short (House) | بيت القصير | Bayt al-Qaṣīr |
| Short al-Bayth | قصير البيث | Qasīr al-Bayth |
| Short Stone | حجر قصير | Ḥajr Qaṣīr |
| Šibenik (City) | صبناجي | Ṣibnājī |
| Sicily (Island) | صقلية | Ṣiqiliya |
| Sigus | تينجس | Tīnjas |
| Sila (River) | الصيلا | al-Ṣīlā |
| Simeri (River) | سيمري | Sīmrī |
| Simeri (Town and River) | سلميرة | Salmīra |
| Simeto (River) | وادي موسى | Wādī Mūsa |
| Sinan (Fort) | سنان | Sinān |
| Sinni (River) | سكنة | Sakna |
| Siponto / Vieste (City) | سيبنت / باستية / بستية | Sībint / Bāstiyya / Bastiyya |
| Sirino (Mountain) | سرين | Sarīn |
| Sirte (City) | صرت | Ṣirt |
| Skanes (Fort) | شقانس | Shaqānis |
| Solanto (Port) | الشبكة | al-Shabaka |
| Soleto | سليت | Salīt |
| Sorrento (City) | سرنت | Sarant |
| Sounine (Fort) | صونين | Ṣūnīn |

| Our Translation | Arabic Name | IJMES Transliteration |
|---|---|---|
| Sousse (City) | سوسة | Sūsa |
| Sperlinga (Stronghold/Fortress) | اسبرلكنة | Isbrilikna |
| Split (City) | اسبالطو | Isbālṭū |
| Spring of the Reeds (Port) | عين القصب | ʿAin al-Qaṣb |
| Spring of the Sultan (Port) | عين السلطان | ʿAin al-Sulṭān |
| Stabia (Port) | اسطابة | Isṭāba |
| Step (Port) | الدرجة | al-Daraja |
| Stilo | استيلو/اسطيلو | Istīlū/ Isṭīlū |
| Ston (City) | شثغنو | Shathghanū |
| Stone of ʿAmār | حجر عمار | Ḥijr ʿAmār |
| Strait Connected to the Sea of Darkness | الزقاق المتصل ببحر الظلمات | al-Zuqāq al-Mutaṣil bi-Bahr al-Ẓulamāt |
| Stromboli (Island) | استرنجلي / اسطرنجلي | Istranjlī / Isṭranjlī |
| Sutera (Estate) | سطير | Suṭīr |
| Syracuse (City) | سرقوسة | Sarqūsa |
| Tabarka (City) | طبرقة | Ṭabarqa |
| Ṭābla | طابلة | Ṭābla |
| Tacina (Town and River) | طاجنو | Ṭājnū |
| Tafanat | تافنات | Tāfnāt |
| Tailor (Fort) | الخياط | al-Khayyāṭ |
| Tamadafit | تامدفيت | Tāmdafīt |
| Tamadit (City) | تامديت | Tāmadīt |
| Tamajina (Village) | طامجنة | Ṭāmajina |
| Tamasit (Village) | تامسيت | Tāmasīt |
| Tanagro (River) | أديانة | Adyāna |
| Tanidha (Fort) | تنيذة | Tanīdha |
| Taormina (City) | طبرمين | Ṭabramīn |
| Taranto (City) | طارنت | Ṭārant |
| Tarsha Dawud (Fort) | ترشة داود | Tarsha Dawud |
| Tarshish (City) | ترشيش | Tarshīsh |
| Taurasi | طرش | Ṭarash |

| Our Translation | Arabic Name | IJMES Transliteration |
|---|---|---|
| Taverna | طبرنة | Ṭabirna |
| Tavi (Castle) | طابس | Ṭābis |
| Tawergha | تاورغا | Tawarghā |
| Teggiano (Castle) | ديانة | Diyāna |
| Termini (Stronghold/Fortress and River) | ثرمة | Thirma |
| Termoli | ترملس | Tirmalis |
| Terracina (City) | طرجينة | Ṭarjīna |
| Terrasini (Port) | ساقية جنش | Sāqyat Jinsh |
| Tindari | دندارة | Dandāra |
| The Idol (Estate) | ألصنم | al-Ṣanam |
| The Idols (Estate) | الاصنام | al-Aṣnām |
| The Lighthouse (Fort) | المنار | al-Manār |
| Tiaret | تاهرت | Tāhirt |
| Tifash (City) | تيفاش | Tīfāsh |
| Tinja (City) | تينجة | Tīnja |
| Tobna | طبنة | Ṭubna |
| Torchiara | ترجل | Tarjal |
| Torre Cavallo (Cape) | ناورة | Nawra |
| Torre Chianca | كنكا | Kankā |
| Torre delle Pietre Rosse | شنت نقولة ببترة | Shant Niqūla Bibitra |
| Torre S. Giovanni (Port) | أنبانة | Anbāna |
| Torre San Gennaro | شنت جنار | Shant Jinār |
| Torre Santo Spirito | اجيلوا / اشيلوا | Ajīlūwā / Ishīlūwā |
| Tozeur (City) | توزر | Tūzir |
| Trabia | التربيعة | al-Tarbīʿa |
| Trani (City) | اطرانة / طرانة | Iṭrāna / Ṭarāna |
| Trapani (City) | طرابنش | Ṭarābansh |
| Tre Fontane (Port) | عيون عباس | ʿAyūn ʿAbās |
| Tremestieri (Port) | الثلاث كنائس | al-Thalāth Kināʾis |
| Tricarico (City) | ترغارقو / اتريغارقو / اتريكركو | Targhārqū / Itrīghārqū / Itrīkirkū |
| Tripi (Stronghold/Fortress) | بربلس | Barbālis |

| Our Translation | Arabic Name | IJMES Transliteration |
|---|---|---|
| Tripoli (City) | اطرابلس | Aṭirābulus |
| Trogir (City) | ترغورون | Targhūrūn |
| Troina (Castle and River) | طرجينس | Ṭarjīns |
| Tronto (City) | طرنت / ترنت / ترانية | Ṭrant / Trant /Tirāniyya |
| Tropea (City) | أتربية | Atrabiyya |
| Tropea (City) | اتربية | Atrabiya |
| Tuccio | توجش | Tawjish |
| Tuccio | توجس | Tūjis |
| Tufara | بطفارة | Biṭfāra |
| Tufo (Castle) | طوفو / طف | Ṭūfū / Ṭuf |
| Tunis (City) | تونس | Tūnis |
| Tursi | ترسة | Tursa |
| Tusa (Castle and Port) | طزعة | Ṭazʿa |
| Tusihan (Fort) | توسيهان | Tūsīhān |
| Tyrrhenian (Sea) | طرانه | Ṭarāna |
| Ulcinj (City) | ذلوجية | Dhalūjiyya |
| Umag | موغاو / أوماغو | Mūghaw / Awmāghū |
| Umbriatico | ابرياتقو | Ibriyātqū |
| Ustica (Island) | اشتقة | Ashtiqa |
| Valentino (Bridge) | بالنتين | Bālintīn |
| Valleoscura (Caste) | بالسقورة | Bālsqūra |
| Vasto | باستو | Bāstū |
| Vaticano (Cape) | باتقانو / قامو | Bātqānū / Qāmū |
| Vendicari (Port) | دخلة ابن دكني | Dakhlat Ibn Daknī |
| Venerable (Body of Water) | وقور | Wuqūr |
| Venosa (City) | بنوصة | Binūṣa |
| Ventotene (Island) | بنت برة | Bant Bara |
| Vergine Maria (Port) | برقة | Barqa |
| Vesuvius (Mountain) | جبل النار | Jabal al-Nār |
| Vicari (Castle) | بيقوا | Bīqwā |
| Vietri (River/Locality) | باذروا | Bādhrwā |
| Viggiano | بجنال | Bijnāl |

| Our Translation | Arabic Name | IJMES Transliteration |
|---|---|---|
| Villetta Barrea (Castle) | براي | Brāy |
| Vizzini | بزيني | Bizīnī |
| Vlorë (City) | لبلونة | Lablūna |
| Volturno (River) | قبوة | Qabwa |
| Vonitsa (City) | بندسة | Bandasa |
| Vonitsa (City) | يانينا | Yānina |
| Vulcanello (Island) | جبل البركان | Jabal al-Burkān |
| Vulcano (Island) | جزيرة البركان | Jazīrat al-Burkān |
| Vranja (City) | برونة / برانة | Barūna / Barāna |
| Waddan (Lands) | ودان | Waddān |
| Wangara (Lands) | ونقارة | Wanqāra |
| Wardasa | ورداسا | Wardāsā |
| Woods (Port) | الشعراء | al-Shaʿrāʾ |
| Zadar (City) | جاذرة | Jādhra |
| Zaghouan (Mountain) | زغوان | Zaghuwan |
| Zajuna (Fort) | زجونة | Zajūna |
| Zanata (Well) | زناتة | Zanāta |
| Zante (Island) | جاجنت | Jājant |
| Zarat (City) | زرود | Zarud |
| Zarzis (Island) | زيزوا | Zīzuwa |
| Zaton (City) | صنطو / صاطو | Ṣanṭū / Ṣāṭū |
| Zawila (City and Suburb) | زويلة | Zawīla |
| Zeffirio (Cape) | جفيرة | Jafīra |
| Zembra (Island) | الجامور الكبير | al-Jāmūr al-Kabīr |
| Zembretta (Island) | الجامور الصغير | al-Jāmūr al-Saghīr |
| Ziyad (Fort) | زياد | Ziyad |
| Ziyad (Spring) | زياد | Ziyad |

Printed and bound by CPI Group (UK) Ltd, Croydon, CR0 4YY

13/09/2026

14961066-0003